AF410704

Carotenoids in Fresh and Processed Food: Between Biosynthesis and Degradation

Carotenoids in Fresh and Processed Food: Between Biosynthesis and Degradation

Editor

Pasquale Crupi

MDPI • Basel • Beijing • Wuhan • Barcelona • Belgrade • Manchester • Tokyo • Cluj • Tianjin

Editor
Pasquale Crupi
University of Bari "Aldo
Moro"
Italy

Editorial Office
MDPI
St. Alban-Anlage 66
4052 Basel, Switzerland

This is a reprint of articles from the Special Issue published online in the open access journal *Applied Sciences* (ISSN 2076-3417) (available at: http://www.mdpi.com).

For citation purposes, cite each article independently as indicated on the article page online and as indicated below:

LastName, A.A.; LastName, B.B.; LastName, C.C. Article Title. *Journal Name* **Year**, *Volume Number*, Page Range.

ISBN 978-3-0365-3817-4 (Hbk)
ISBN 978-3-0365-3818-1 (PDF)

Cover image courtesy of Pasquale Crupi

© 2022 by the authors. Articles in this book are Open Access and distributed under the Creative Commons Attribution (CC BY) license, which allows users to download, copy and build upon published articles, as long as the author and publisher are properly credited, which ensures maximum dissemination and a wider impact of our publications.
The book as a whole is distributed by MDPI under the terms and conditions of the Creative Commons license CC BY-NC-ND.

Contents

About the Editor

Pasquale Crupi is currently Researcher at the Interdisciplinary Department of Medicine—University of Bari "Aldo Moro". His research interests are mainly focused on metabolomic analyses of food products; in particular, metabolic fingerprinting and metabolite profiling analyses by high resolution and high-throughput technologies (GC–MS, HPLC–MSn, HPLC–MS/MS, NMR) of primary and secondary metabolites (sugars, amino acids, organic acids, polyphenols, carotenoids, aromatic compounds, vitamins, etc.) present in fruit and vegetables (such as grape, cherry, carob, artichoke, olive etc.) and biotransformed products (such as juice, extra virgin olive oil, and wine). He is coauthor of more than 60 publications in international peer-reviewed journals.

Preface to "Carotenoids in Fresh and Processed Food: Between Biosynthesis and Degradation"

Nowadays, consumers look at food not only as a source of energy and nutrition but also an affordable way to promote health and prevent future diseases.

In this context, studying the qualitative and quantitative profiles of natural compounds such as carotenoids, which have antioxidant and anti-inflammatory properties, in food is very relevant. In the last decades, many studies have demonstrated the importance of a diet rich in carotenoids in lowering the onset of certain diseases, such as for numerous types of cancer, cardiovascular diseases, age-related macular degeneration, etc.

Regardless, new insights in this research field are still necessary. Therefore, this *Applied Sciences* Special Issue has collected relevant contributions on interesting aspects related to the composition pattern, biosynthesis and degradation, and overall chemical properties of carotenoids in fresh and processed food, which can improve knowledge in the sector of food science and food chemistry. In this sense, it is worth pointing out that the gathered manuscripts represent valuable advancements in the study of carotenoids.

Of course, naturally thank all authors contributing to this Special Issue in *Applied Sciences* for their scientific input and experimental efforts throughout the project.

Pasquale Crupi
Editor

Editorial

Carotenoids in Fresh and Processed Food: Between Biosynthesis and Degradation

Pasquale Crupi

Dipartimento Interdisciplinare di Medicina, Università degli Studi Aldo Moro Bari, P.zza Giulio Cesare, 70124 Bari, Italy; pasquale.crupi@uniba.it; Tel.: +39-347-125-2849

Citation: Crupi, P. Carotenoids in Fresh and Processed Food: Between Biosynthesis and Degradation. *Appl. Sci.* **2022**, *12*, 1689. https://doi.org/10.3390/app12031689

Received: 22 January 2022
Accepted: 3 February 2022
Published: 7 February 2022

Publisher's Note: MDPI stays neutral with regard to jurisdictional claims in published maps and institutional affiliations.

Copyright: © 2022 by the author. Licensee MDPI, Basel, Switzerland. This article is an open access article distributed under the terms and conditions of the Creative Commons Attribution (CC BY) license (https://creativecommons.org/licenses/by/4.0/).

1. Introduction

Currently, there is a general trend in food science to link food and health in line with consumers' concern about what is in their food and how what they eat can promote well-being. Thus, food is considered today not only a source of energy but also an affordable way to prevent future diseases. In this context, studying carotenoids content in food is very relevant. Indeed, epidemiological studies have demonstrated that the consumption of diets rich in carotenoids is associated with a lower incidence of cancer, cardiovascular diseases, and age-related macular degeneration, mainly due to their antioxidant and provitamin A activity [1]. Although many works have been conducted concerning the presence and properties of carotenoids in food [2], some challenges must be still faced in this research field: The role of carotenoids as antioxidants and its mechanism of action need to be investigated further; detailed qualitative and quantitative composition of carotenoids in underutilized fruits and vegetables is required in order to contribute significant information to select nutrient rich plants for food formulation; how emerging packaging and processing techniques (i.e., high electric field pulse, high-pressure CO_2, etc.) can preserve the content of carotenoids in processed food products needs to be understood; the complete understanding of carotenoid biosynthesis, regulation, and roles of various carotenoid derivatives for edible plants and animals is still not well established; and detailed studies for identifying the pre- and post-harvesting favorable factors (i.e., elicitors, cooking methods, etc.), which improve the bioavailability and bioaccessibility of carotenoids from different foods, are necessary.

The Special Issue "Carotenoids in Fresh and Processed Food: Between Biosynthesis and Degradation" was aimed to invite worldwide scholars (particularly experts in the sector of food science and food chemistry) to submit their most interesting communications, reviews, and original articles that can improve the knowledge in the field of carotenoids in food.

Potential topics included, but were not restricted to, carotenoids and apocarotenoids chemistry and biosynthesis, structural isomerization and degradation, content in vegetable and non-vegetable foods, and bioavailability and bioaccesibility methods of analysis.

2. Carotenoids in Fresh and Processed Food: Between Biosynthesis and Degradation

The aim of this Special Issue was to group the most recent and relevant research in relation to the aforementioned topics regarding carotenoids in food into a single document. Subsequently, the possibility of publishing a book with the contributions of all authors has been assessed. There were six papers submitted to this Special Issue, and five of them were accepted. In the following paragraphs, a summary of these papers with their most relevant findings is presented.

The first paper [3] deals with the protection of β-Carotene from photodegradation. The authors of this work showed how β-Carotene degrades rapidly in a 2% oil-in-water emulsion, made from food-grade soy oil with 7.4 mg β-carotene/mL oil, during storage and when exposed to light. However, the addition of clove oil (2.0, 4.0, or 8.0 μL/mL of emulsion) prevented the photodegradation of β-carotene, regardless of the ratio between clove oil and β-carotene in the concentration range studied. Since plant phenols have been demonstrated to efficiently regenerate carotenoids from their initial photooxidation products, the authors concluded that the observed regeneration of β-carotene was due to eugenol, the main plant phenol of clove oil to occur in the oil-water interface. Therefore, clove oil in low concentrations may find use as a natural protectant of provitamin A in enriched foods during retail display.

The second paper [4] presented how two typical plant hormones, namely salicylic acid (SA) and methyl jasmonate (MeJA), were able to regulate the accumulation of flavonoids (i.e., eriocitrin, narirutin, and poncirin) and carotenoids (i.e., β-cryptoxanthin) in the juice sacs of Satsuma mandarin in vitro. The results showed that SA treatment was effective in enhancing the contents of eriocitrin, narirutin, poncirin, and β-cryptoxanthin, whilst MeJA treatment inhibited these compounds accumulation in the juice sacs ($p < 0.05$). Moreover, gene expression analysis confirmed that the changes of flavonoid and carotenoid contents were highly regulated at the transcriptional level. In particular, a transcriptional factor CitWRKY70 was identified in the microarray analysis, which was induced by the SA treatment while being suppressed by the MeJA treatment. Since the change in the expression of CitWRKY70 was consistent with that of flavonoid and carotenoid biosynthetic key genes, this finding indicated that CitWRKY70 might be involved in the regulation of the investigated compounds content in the juice sacs of citrus fruit in response to SA and MeJA treatments.

In the third article, Gałązka-Czarnecka et al. [5] studied the influence of light at different wavelengths (white light at 380–780 nm, UVA at 340 nm, blue light at 440 nm, and red light at 630 nm) and pulsed electric field (PEF) at different strength (1, 2.5 and 5 kV/cm) on the content of carotenoids (i.e., lutein, zeaxanthin, and β-carotene) in red clover sprouts. The experiment was carried out in a climatic chamber with phytotron system under seven growing conditions differing in light-emitting diode (LED) wavelengths and PEF strength applied before sowing. Lutein was found as the dominant carotenoid in germinating red clover seeds, with content varying from 743 mg/kg in sprouts grown in red light to 888 mg/kg in sprouts grown in blue light. Blue light treatment during the red clover sprouts growing had the most beneficial effect in enhancing carotenoids content up to 42% in β-carotene, 19% in lutein, and 14% in zeaxanthin. An increase of β-carotene (8.5%) and lutein (6%) amount was also obtained with white light without PEF pre-treatment; conversely zeaxanthin decreased by about 3.3%. Therefore, the authors concluded that PEF pre-treatment may increase mainly the content of β-carotene in red clover sprouts.

The presence of carotenoids in grape berries is well documented [6]; the grape variety and viticulture practices, but also climate conditions and geographic origin, can influence their qualitative and quantitative profile as well as their degradation during grape ripening from véraison to harvest [7]. The last two works, belonging to this SI, treated about effective practices for conditioning carotenoids degradation in grapes. In particular, Asproudi et al. [8] investigated the impact of bunch microclimate on the evolution of some relevant carotenoids (i.e., neoxanthin, luteinin, and β-carotene) in Nebbiolo grapes, collected from green phase up to harvest, during two consecutive seasons. Overall, higher temperature in the less vigorous and south facing vineyards led to lower amounts of carotenoids, both during ripening and at harvest. Lutein and neoxanthin contents (μg/berry) varied similarly in both seasons and achieved a maximum after veraison, especially in the cooler plots. Therefore, a variety effect on the lutein seasonal trend was hypothesized. Conversely, b-carotene content remained generally constant during ripening, with the exception of the south plots showing dissimilar evolution between the seasons. This observation allowed the authors to conclude that bunch zone temperature and light condition may affect both synthesis and degradation of grape carotenoids determining their amount and profile at harvest.

Crupi et al. [9] aimed to study the effect of the foliar application of yeast extracts (YE) to Negro Amaro and Primitivo grapevines on the carotenoid content during grape ripening and the difference between the resulting véraison and maturity (ΔC). The results showed that β-carotene and (allE)-lutein were the most abundant carotenoids in all samples, ranging from 60% to 70% of total compounds. Their levels, as well as those of violaxanthin, (9-Z)-neoxanthin, and 5,6-epoxylutein, decreased during ripening. This was especially observed in treated grapes, with ΔC values from 2.6 to 4.2-fold higher than in untreated grapes. Thereby, the YE treatment has proved to be effective in improving the C_{13}-norisoprenoid aroma potentiality of Negro Amaro and Primitivo, which are fundamental cultivars in the context of Italian wine production.

Funding: This research received no external funding.

Institutional Review Board Statement: Not applicable.

Informed Consent Statement: Not applicable.

Data Availability Statement: Not applicable.

Conflicts of Interest: The author declares no conflict of interest.

References

1. Eggersdorfer, M.; Wyss, A. Carotenoids in human nutrition and health. *Arch. Biochem. Biophys.* **2018**, *652*, 18–26. [CrossRef] [PubMed]
2. Meléndez-Martínez, A.J.; Mandić, A.I.; Bantis, F.; Böhm, V.; Borge, G.I.A.; Brnčić, M.; Bysted, A.; Cano, M.P.; Dias, M.G.; Elgersma, A.; et al. A comprehensive review on carotenoids in foods and feeds: Status quo, applications, patents, and research needs. *Crit. Rev. Food Sci. Nutr.* **2021**, *5*, 1–51. [CrossRef] [PubMed]
3. Zhou, Y.-M.; Chang, H.-T.; Zhang, J.-P.; Skibsted, L.H. Clove Oil Protects β-Carotene in Oil-in-Water Emulsion against Photodegradation. *Appl. Sci.* **2021**, *11*, 2667. [CrossRef]
4. Yamamoto, R.; Ma, G.; Zhang, L.; Hirai, M.; Yahata, M.; Yamawaki, K.; Shimada, T.; Fujii, H.; Endo, T.; Kato, M. Effects of Salicylic Acid and Methyl Jasmonate Treatments on Flavonoid and Carotenoid Accumulation in the Juice Sacs of Satsuma Mandarin In Vitro. *Appl. Sci.* **2020**, *10*, 8916. [CrossRef]
5. Gałązka-Czarnecka, I.; Korzeniewska, E.; Czarnecki, A.; Kiełbasa, P.; Dróżdż, T. Modelling of Carotenoids Content in Red Clover Sprouts Using Light of Different Wavelength and Pulsed Electric Field. *Appl. Sci.* **2020**, *10*, 4143. [CrossRef]
6. Crupi, P.; Coletta, A.; Milella, R.A.; Palmisano, G.; Baiano, A.; La Notte, E.; Antonacci, D. Carotenoid and Chlorophyll derived compounds in some wine grapes grown in Apulian region. *J. Food Sci.* **2010**, *75*, S191–S198. [CrossRef] [PubMed]
7. Crupi, P.; Coletta, A.; Antonacci, D. Analysis of carotenoids in grapes to predict norisoprenoid varietal aroma of wines from Apulia. *J. Agric. Food Chem.* **2010**, *58*, 9647–9656. [CrossRef] [PubMed]
8. Asproudi, A.; Petrozziello, M.; Cavalletto, S.; Ferrandino, A.; Mania, E.; Guidoni, S. Bunch Microclimate Affects Carotenoids Evolution in cv. Nebbiolo (*V. vinifera* L.). *Appl. Sci.* **2020**, *10*, 3846. [CrossRef]
9. Crupi, P.; Santamaria, M.; Vallejo, F.; Tomás-Barberán, F.A.; Masi, G.; Caputo, A.R.; Battista, F.; Tarricone, L. How Pre-Harvest Inactivated Yeast Treatment May Influence the Norisoprenoid Aroma Potential in Wine Grapes. *Appl. Sci.* **2020**, *10*, 3369. [CrossRef]

Communication

Clove Oil Protects β-Carotene in Oil-in-Water Emulsion against Photodegradation

Yi-Ming Zhou [1], Hui-Ting Chang [1], Jian-Ping Zhang [1,*] and Leif H. Skibsted [2,*]

[1] Department of Chemistry, Renmin University of China, Beijing 100872, China; zym1990@ruc.edu.cn (Y.-M.Z.); htchang1991@163.com (H.-T.C.)

[2] Department of Food Science, University of Copenhagen, Rolighedsvej 30, DK-1958 Frederiksberg C, Denmark

* Correspondence: jpzhang@ruc.edu.cn (J.-P.Z.); ls@food.ku.dk (L.H.S.); Tel.: +86-10-62516604 (J.-P.Z.); +45-3533-3221 (L.H.S.); Fax: +86-10-62516444 (J.-P.Z.)

Abstract: β-Carotene degrades rapidly in a 2% oil-in-water emulsion, made from food-grade soy oil with 7.4 mg β-carotene/mL oil, during storage and when exposed to light. Added clove oil (2.0, 4.0, or 8.0 μL/mL of emulsion) protects against the photodegradation of β-carotene, regardless of the ratio between clove oil and β-carotene in the concentration range studied, suggesting that the regeneration of β-carotene is caused by eugenol, the principal plant phenol of clove oil to occur in the oil-water interface. Therefore, clove oil in low concentrations may find use as a natural protectant of provitamin A in enriched foods during retail display.

Keywords: eugenol; photoprotection; provitamin A

Citation: Zhou, Y.-M.; Chang, H.-T.; Zhang, J.-P.; Skibsted, L.H. Clove Oil Protects β-Carotene in Oil-in-Water Emulsion against Photodegradation. *Appl. Sci.* **2021**, *11*, 2667. https://doi.org/10.3390/app11062667

Academic Editor: Pasquale Crupi

Received: 19 February 2021
Accepted: 13 March 2021
Published: 17 March 2021

Publisher's Note: MDPI stays neutral with regard to jurisdictional claims in published maps and institutional affiliations.

Copyright: © 2021 by the authors. Licensee MDPI, Basel, Switzerland. This article is an open access article distributed under the terms and conditions of the Creative Commons Attribution (CC BY) license (https://creativecommons.org/licenses/by/4.0/).

1. Introduction

While β-Carotene (β-Car) is an important provitamin A, it is sensitive to light and degrades rapidly in plant oils or in plant oil emulsions during storage under ambient conditions [1–4]. Vitamin A deficiency is a major challenge worldwide, especially for children's nutrition, and urgently calls for practical solutions [5,6].

Recently, it was found that plant phenols regenerate β-Car and other carotenoids (Car) from their initial photooxidation product, the carotenoid radical cations (Car$^{\bullet+}$), through electron transfer from the reducing phenol group φ-OH [7]:

$$Car^{\bullet+} + \varphi\text{-OH} \rightarrow Car + \varphi\text{-O}^{\bullet} + H^+ \tag{1}$$

The regeneration of β-Car corresponding to the reaction of Equation (1) was surprisingly found to be the most efficient for moderately reducing plant phenols, such as eugenol, while strongly reducing plant phenols, like tea catechins, showed no regeneration of β-Car, but displayed enhanced photobleaching [7–9].

Eugenol and isoeugenol, the main constituents of clove oil [10], are moderately reducing plant phenols that have been found to regenerate β-Car efficiently from the radical cation formed by photolysis of β-Car. This reduction occurs in alkaline chloroform/methanol as an electron-withdrawing solvent [11]. The ordering of the anions of the plant phenols according to the rate of regeneration of carotenoids could further be accounted for by the Marcus theory of electron transfer [12]. According to this theory, the maximal rate of electron transfer corresponds to a driving force matching the reorganization energy in the transition state for electron transfer. Notably, for a larger driving force, the rate of electron transfer enters the so-called inverted region with a higher activation barrier, and accordingly, lower rates are seen for quercetin and tea catechins [7,12].

The more practical aspects of the Marcus theory for electron transfer have not yet been exploited in relation to food preservation. However, the protection of β-Car, as a provitamin A in an oil-in-water emulsion in a functional food, could provide a proof of concept, where the use of Marcus theory could be moved from model systems involving chlorinated solvents of high pH into a real food system. Accordingly, clove oil, with high content of moderately reducing plant phenols and its worldwide use in food and beverages, was combined with β-Car in an oil-in-water emulsion and stored under illumination in ambient conditions with the objective of protecting provitamin A against degradation during retail display. The present study aimed to explore whether the Marcus theory for electron transfer could be used to design optimal protection of a light-sensitive vitamin.

2. Materials and Methods

2.1. Materials

All-*trans*-β-carotene (β-Car) was from Sigma-Aldrich (St. Louis, MO, USA). Clove oil, containing 85% eugenol and isoeugenol, was from O'plants (Shanghai, China). Soybean oil was from Yihai Kerry Food Co., Ltd. (Beijing, China). Whey protein isolates (WPI) were from HIRMAR (Los Angeles, CA, USA). Lecithin (95%) was from Arbor Star Biological Technology Co., Ltd. (Beijing, China).

2.2. Preparation of Emulsion

β-Car (40 mg, accurately weighted) and lecithin (2400 mg) were mixed in 5.4 mL soybean oil and stirred at 1000 rpm for 3 h in the dark to fully dissolve the mixture in the oil phase. The water phase contained 12 g WPI in 280 mL deionized water with the pH of the aqueous solution adjusted to 7.0. This solution was adjusted by dropwise addition of dilute HCl and NaOH, while pH was monitored electrochemically. The emulsion was prepared by mixing the oil phase with the aqueous solution and homogenizing the mixture at 13,000 rpm for 5 min using an FA25 homogenizer (Shanghai, China). Subsequently, 10 mL of emulsion samples were added to glass jars before adding the clove oil (0.02 mL, 0.04 mL, or 0.08 mL) to different samples, which were homogenized at 13,000 rpm for 2 min. In this study, 10 mL emulsion without clove oil served as the control sample. The emulsion had a fat content of approximately 2%, which is comparable to milk and other nutritive beverages. The final concentration of eugenol and isoeugenol from clove oil was 1.7, 3.4, or 6.8 µL/mL emulsion. All samples were stored under light (spectral distribution in the 300−800 nm range, 22,000 Lx warm white similar to light used for illumination during retail display) at 25 °C. Control emulsion samples were stored in the dark at 25 °C. The main experiment, as described in Figures 1 and 2 as well as in Table 1, was in storage for three weeks. The standard deviation of each color measurement was less than 1%.

Figure 1. Appearance of oil-in-water with β-Car during storage under ambient conditions and on exposure to light in glass jars with and without addition of 4 µL/mL emulsion of clove oil.

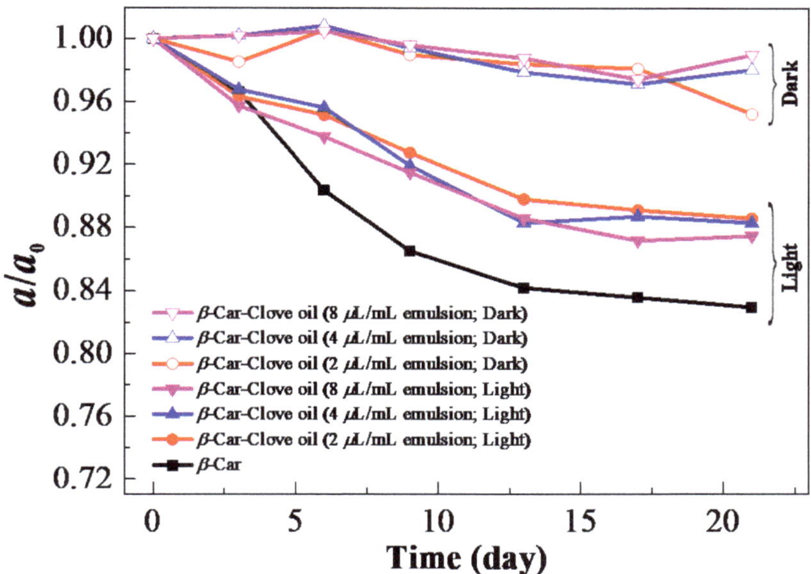

Figure 2. Red bleaching (relative redness a/a_0) of oil-in-water emulsion with β-Car with or without the addition of clove oil during storage in the dark or exposed to light of 22,000 Lx at 25 °C.

Table 1. The redness parameter a of β-Car and β-Car-clove oil on light exposure compared to dark storage for different days. The redness parameter of 0 day is defined as a_0.

Sample / Day	0	3	6	9	13	17	21
β-Car	24.12	23.31	21.80	20.87	20.31	20.16	20.01
β-Car-clove oil (2 μL/mL emulsion; Light)	24.69	23.78	23.49	22.90	22.17	22.00	21.87
β-Car-clove oil (4 μL/mL emulsion; Light)	24.45	23.66	23.38	22.48	21.59	21.69	21.59
β-Car-clove oil (8 μL/mL emulsion; Light)	23.39	22.39	21.93	22.48	20.72	20.39	20.46
β-Car-clove oil (2 μL/mL emulsion; Dark)	24.27	23.91	24.40	24.02	23.88	23.81	23.11
β-Car-clove oil (4 μL/mL emulsion; Dark)	24.07	24.12	24.27	23.93	23.56	23.38	23.60
β-Car-clove oil (8 μL/mL emulsion; Dark)	22.93	22.97	23.04	22.84	22.65	22.34	22.70

2.3. UV-Visible Absorption Spectroscopy

UV-visible absorption spectra were measured on a Cary50 spectrophotometer (Varian Inc., Palo Alto, CA, USA), using 1.0 cm quartz cells. According to Lambert-Beer's law, the soybean oil acted as a mixed low-polarity solvent for the concentration of β-Car as it relates to the absorbance:

$$c = \frac{A_\lambda}{\varepsilon_\lambda \cdot b} \tag{2}$$

Equation (2) was used for quantification in this study [13]. In Equation (2), c is the molar concentration (mol·L^{-1}) of β-Car. A_λ is the measured absorbance, and ε_λ is the molar extinction coefficient of β-Car in pure soybean oil (1.43×10^5 L·mol^{-1}·cm^{-1}) at the maximal absorption wavelength (λ) of 462 nm. b is the optical pathlength of the cuvette (1 cm).

2.4. Color Measurement

The LAB Hunter values of the emulsion samples were measured multiple times by a PR-780 Spectrophotometer (Photo Research, Los Angeles, CA, USA) during storage for up to 21 days. In the main experiment, photographs of the samples were taken regularly during storage using a digital camera. The light source for color measurement was a tungsten lamp (40 W), and a standard, white tile served as a background.

3. Results and Discussion

β-Car is lipophilic ($\log P = 12.2$) and dissolves in the soy oil of the oil-in-water emulsion. Clove oil consists mainly of eugenol with $\log P = 2.49$ [14,15] which distributes between the oil and the aqueous phase. As evident from Figure 1, the emulsion appeared homogeneously red. The concentration of β-Car in the emulsion oil phase was 7.4 mg/mL soy oil, while the phenols from clove oil were distributed between the two phases. Soy oil was selected for the oil phase of the emulsion as it is edible, with good nutritive value, and is available worldwide. Lecithin with 8.4 mg/mL emulsion was also added because it is commonly used as an emulsifier in foods.

When stored in the dark, the color remained constant, as depicted in Figure 2. In Table 1, a, i.e., the redness parameter of the LAB color system, is shown for 21 days of storage, while a_0 is the redness parameter of day 0. The presence of clove oil did not affect the color during dark storage at any of the three concentrations. This finding was in agreement with the robustness toward uncatalyzed degradation of β-Car which was previously observed [16].

Upon exposure to light, the redness faded, as was evident from visual inspection; see Figure 1. Carotenoids are generally sensitive to radiation, including light and γ-irradiation [17]. The redness parameter a also showed a significant decrease during storage when exposed to light (Table 1 and Figure 2). The presence of clove oil clearly provided protection, as bleaching was reduced to approximately half of that in the emulsion without clove oil. Notably, the protection of color, and accordingly, of β-Car, was not dependent on either the amount of clove oil added or the concentration of the plant phenols in the concentration range studied (clove oil between 2.0 μL/mL and 8.0 μL/mL), due to the saturation of plant phenols at the emulsion interface. The decrease of the redness parameter a could be described by a mono-exponential model function for each of the three independent experiments for which the rate constant was 0.089 days^{-1}, and was not dependent on the clove oil concentration. This type of protection was similar to that of plant phenols toward the carotenoids involved in the visual function [18].

The light source used for the storage experiment had an intensity of 22,000 Lx, was mainly in the visible region, and had a minor UV-component. The glass of the jars further served as a UV-filter. As seen from the absorption spectra of Figure 3, the light was absorbed by β-Car rather than by the clove oil phenols. Excitation of β-Car to the singlet or triplet states generated radical cations, leading to bleaching:

$$\mathrm{Car} + h\nu \rightarrow {}^1\mathrm{Car}* \tag{3}$$

$$^1\mathrm{Car}* \rightarrow {}^3\mathrm{Car}* \tag{4}$$

$$^3\mathrm{Car}*/{}^1\mathrm{Car}* \rightarrow \mathrm{Car}^{\bullet+} + e^- \,(\mathrm{solv.}) \tag{5}$$

$$\mathrm{Car}^{\bullet+} \rightarrow \text{Degradation products} \tag{6}$$

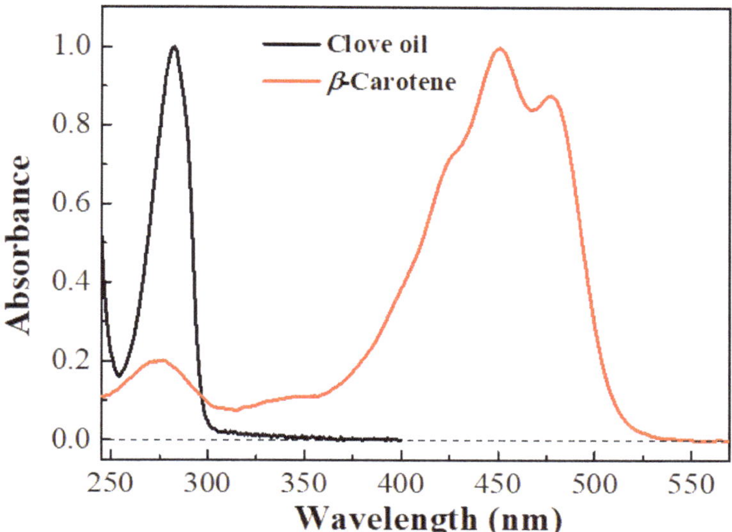

Figure 3. Absorption spectra of β-Car in methanol normalized at 450 nm and of clove oil in methanol normalized at 283 nm.

According to Equations (1) and (6), the regeneration of the carotenoid from radical cations will compete with its degradation. As seen in Figure 2, the bleaching was independent of the concentration of eugenol in the concentration range studied.

The regeneration rate previously found for homogeneous solution [7]:

$$\frac{d[\text{Car}]}{dt} = k_2[Car^+][\varphi\text{–OH}] \tag{7}$$

seems for the present conditions independent of the total phenol concentration in the emulsion. This apparent zero-order dependence on the plant phenol for the emulsion probably indicated: (i) a rapid electron transfer, and (ii) an apparent similar excess of phenol available for reduction under all conditions investigated. These observations pointed toward a mechanism occurring in the emulsion interface that was saturated with the plant phenols. The rate expression of Equation (7) was based on a series of more systematic kinetic studies in homogenous solutions [7–9,11]. The observed kinetics for the photodegradation of β-Car in the oil-in-water emulsion can be accommodated within this theory, including the partition of eugenol between the homogeneous aqueous phase and the heterogeneous oil phase. The distribution between water and oil may be adjusted as eugenol is consumed.

In the oil-in-water emulsion, the protection of β-Car by clove oil is an important finding, since regeneration occurs at neutral pH as compared to the conditions of high pH used in model studies [7–9,11]. The phenols of clove oil and not only their anions are sufficiently reducing for the donation of an electron, and have matching reduction potential according to Marcus' theory to reduce the carotenoid radical cation [12]. Isoeugenol and especially eugenol may be unique in this respect; nevertheless, other plant oils and plant phenols with similar, moderate reduction potentials are now being investigated for their ability to protect carotenoids against light degradation in food.

Designing functional foods with better shelf life is encouraging, as there is a serious problem with vitamin A deficiency worldwide [19]. Moreover, the use of plant oils will provide such products with a natural image of sustainability. The practical application of these findings still needs further development, but in light of the simple procedures required, the perspective seems encouraging, especially for developing countries.

4. Conclusions

Our results show that clove oil protects β-carotene in an oil-in-water emulsion from photodegradation due to the content of moderately reducing plant phenols. It serves as a proof of concept for the use of the Marcus theory for electron transfer as a strategy for the protection of vitamin A and provitamin A compounds, thus addressing a global problem.

Author Contributions: Conceptualization of the study and wrote the manuscript, J.-P.Z. and L.H.S.; designed experiments, performed experiments and analysis of data, Y.-M.Z. and H.-T.C. All authors have read and agreed to the published version of the manuscript.

Funding: This work has been supported by the Natural Science Foundation of China (Grant Nos. 21673288, 21573284).

Institutional Review Board Statement: Not applicable.

Informed Consent Statement: Not applicable.

Data Availability Statement: The data and analyses from the current study are available from the corresponding authors upon reasonable request.

Conflicts of Interest: The authors declare no conflict of interest.

References

1. Bohm, F.; Edge, R.; Truscott, G. Interactions of dietary carotenoids with activated (singlet) oxygen and free radicals: Potential effects for human health. *Mol. Nutr. Food Res.* **2012**, *56*, 205–216. [CrossRef] [PubMed]
2. Martínez, A.; Vargas, R.; Galano, A. What is important to prevent oxidative stress? A theoretical study on electron-transfer reactions between carotenoids and free radicals. *J. Phys. Chem. B* **2009**, *113*, 12113–12120. [CrossRef] [PubMed]
3. Schroeder, M.T.; Becher, E.M.; Skibsted, L.H. Molecular mechanism of antioxidant synergism of tocotrienols and carotenoids in palm oil. *J. Agric. Food Chem.* **2006**, *54*, 3445–3453. [CrossRef] [PubMed]
4. Zhao, D.; Yu, D.; Kim, M.; Gu, M.Y.; Kim, S.M.; Pan, C.H.; Kim, G.H.; Chung, D.J.F.C. Effects of temperature, light, and pH on the stability of fucoxanthin in an oil-in-water emulsion. *Food Chem.* **2019**, *291*, 87–93. [CrossRef] [PubMed]
5. Akhtar, S.; Ahmed, A.; Randhawa, M.A.; Atukorala, S.; Arlappa, N.; Ismail, T.; Ali, Z. Prevalence of vitamin a deficiency in south asia: Causes, outcomes, and possible remedies. *J. Health Popul. Nutr.* **2013**, *4*, 413–423. [CrossRef] [PubMed]
6. Rodriguez-Amaya, D.B. Status of carotenoid analytical methods and in vitro assays for the assessment of food quality and health effects. *Curr. Opin. Food Sci.* **2015**, *1*, 56–63. [CrossRef]
7. Cheng, H.; Han, R.M.; Zhang, J.P.; Skibsted, L.H. Electron transfer from plant phenolates to carotenoid radical cations. Antioxidant interaction entering the Marcus theory inverted region. *J. Agric. Food Chem.* **2014**, *62*, 942–949. [CrossRef] [PubMed]
8. Song, L.L.; Liang, R.; Li, D.D.; Xing, Y.D.; Han, R.M.; Zhang, J.P.; Skibsted, L.H. Beta-carotene radical cation addition to green tea polyphenols. Mechanism of antioxidant antagonism in peroxidizing liposomes. *J. Agric. Food Chem.* **2011**, *59*, 12643–12651. [CrossRef] [PubMed]
9. Huvaere, K.; Skibsted, L.H. Flavonoids protecting food and beverages against light. *J. Sci. Food Agric.* **2015**, *95*, 20–35. [CrossRef] [PubMed]
10. Mihara, S.; Shibamoto, T. Photochemical reactions of eugenol and related compounds: Synthesis of new flavor chemicals. *J. Agric. Food Chem.* **1982**, *30*, 1982–1985. [CrossRef]
11. Chang, H.T.; Cheng, H.; Han, R.M.; Wang, P.; Zhang, J.P.; Skibsted, L.H. Regeneration of beta-carotene from radical cation by eugenol, isoeugenol, and clove oil in the marcus theory inverted region for electron transfer. *J. Agric. Food Chem.* **2017**, *65*, 908–912. [CrossRef] [PubMed]
12. Marcus, R.A. Electron transfer reactions in chemistry. theory and experiment. *Rev. Mod. Phys.* **1993**, *65*, 599–610. [CrossRef]
13. Biehler, E.; Mayer, F.; Hoffmann, L.; Krause, E.; Bohn, T. Comparison of 3 spectrophotometric methods for carotenoid determination in frequently consumed fruits and vegetables. *J. Food Sci.* **2010**, *75*, 55–61. [CrossRef] [PubMed]
14. De Cassia, S.S.R.; Pereira, M.M.; Freire, M.G.; Coutinho, J.A.P. Evaluation of the effect of ionic liquids as adjuvants in polymer-based aqueous biphasic systems using biomolecules as molecular probes. *Sep. Purif. Technol.* **2018**, *196*, 244–253.

15. Taherpour, A.A.; Taherpour, A.; Taherpour, Z.; Taherpour, O. Relationship study of octanol–water partitioning coefficients and total biodegradation of linear simple conjugated polyene and carotene compounds by use of the Randićindex and maximum UV wavelength. *Phys. Chem. Liq.* **2009**, *47*, 349–359. [CrossRef]
16. Mortensen, A.; Skibsted, L.H. Kinetics and Mechanism of the Primary Steps of Degradation of Carotenoids by Acid in Homogeneous Solution. *J. Agric. Food Chem.* **2000**, *48*, 279–286. [CrossRef] [PubMed]
17. Boehm, F.; Edge, R.; Truscott, T.G.; Witt, C. A dramatic effect of oxygen on protection of human cells against γ-radiation by lycopene. *FEBS Lett.* **2016**, *590*, 1086–1093. [CrossRef] [PubMed]
18. Skibsted, L.H. Anthocyanidins regenerating xanthophylls: A quantum mechanical approach to eye health. *Curr. Opin. Food Sci.* **2018**, *20*, 24–29. [CrossRef]
19. Skibsted, L.H. Vitamin and non-vitamin antioxidants and their interaction in food. *J. Food Drug. Anal.* **2012**, *20*, 355–358. [CrossRef]

Article

Effects of Salicylic Acid and Methyl Jasmonate Treatments on Flavonoid and Carotenoid Accumulation in the Juice Sacs of Satsuma Mandarin In Vitro

Risa Yamamoto [1,†], Gang Ma [1,2,†], Lancui Zhang [2], Miki Hirai [2,‡], Masaki Yahata [1,2], Kazuki Yamawaki [1,2], Takehiko Shimada [3], Hiroshi Fujii [3], Tomoko Endo [3] and Masaya Kato [1,2,*]

[1] Graduate School of Integrated Science and Technology, Department of Bioresource Sciences, Faculty of Agriculture, Shizuoka University, 836 Ohya, Suruga, Shizuoka 422-8529, Japan; Yamamoto.risa.15@shizuoka.ac.jp (R.Y.); ma.gang@shizuoka.ac.jp (G.M.); yahata.masaki@shizuoka.ac.jp (M.Y.); yamawaki.kazuki@shizuoka.ac.jp (K.Y.)

[2] Department of Bioresource Sciences, Faculty of Agriculture, Shizuoka University, 836 Ohya, Suruga, Shizuoka 422-8529, Japan; zhang.lan.cui@shizuoka.ac.jp (L.Z.); miki1_hirai@pref.shizuoka.lg.jp (M.H.)

[3] National Institute of Fruit Tree Science (NIFTS), National Agriculture and Bio-Oriented Research Organization (NARO), Shizuoka, Shizuoka 424-0292, Japan; tshimada@affrc.go.jp (T.S.); hfujii@affrc.go.jp (H.F.); tomoen@affrc.go.jp (T.E.)

[*] Correspondence: kato.masaya@shizuoka.ac.jp; Tel.: +81-54-238-4830

[†] These authors contributed equally: Risa Yamamoto and Gang Ma.

[‡] Current address: Shizuoka Prefectural Agriculture and Forestry Research Institute, Fruit Tree Research Center, Mobata, Shimizu-ku, Shizuoka 424-0101, Japan.

Received: 30 October 2020; Accepted: 10 December 2020; Published: 14 December 2020

Abstract: Salicylic acid and jasmonic acid are two important plant hormones that trigger the plant defense responses and regulate the accumulation of bioactive compounds in plants. In the present study, the effects of salicylic acid (SA) and methyl jasmonate (MeJA) on flavonoid and carotenoid accumulation were investigated in the juice sacs of Satsuma mandarin in vitro. The results showed that SA treatment was effective to enhance the contents of eriocitrin, narirutin, poncirin, and β-cryptoxanthin in the juice sacs ($p < 0.05$). In contrast, the MeJA treatment inhibited flavonoid and carotenoid accumulation in the juice sacs ($p < 0.05$). Gene expression results showed that the changes of flavonoid and carotenoid contents in the SA and MeJA treatments were highly regulated at the transcriptional level. In addition, a transcriptional factor *CitWRKY70* was identified in the microarray analysis, which was induced by the SA treatment, while suppressed by the MeJA treatment. In the SA and MeJA treatments, the change in the expression of *CitWRKY70* was consistent with that of flavonoid and carotenoid biosynthetic key genes. These results indicated that *CitWRKY70* might be involved in the regulation of flavonoid and carotenoid accumulation in response to SA and MeJA treatments in the juice sacs of citrus fruit.

Keywords: flavonoid; carotenoid; salicylic acid; methyl jasmonate; citrus fruit; juice sacs

---

1. Introduction

Fruit ripening is a complex developmental process involving a number of physiological and biochemical changes that are thought to be under hormonal and environmental control [1,2]. Carotenoids, whose accumulation increase during ripening in citrus, are secondary metabolites with antioxidant effects in plants and are an important group of natural pigments. In particular, β-cryptoxanthin is an important xanthophyll accumulated in fruits and vegetables. Some epidemiological studies have reported that dietary intake of β-cryptoxanthin reduced the risks of certain diseases, especially cancers, diabetes, and rheumatism, because of its antioxidant activity [3–8].

In citrus fruit, flavonoids are also important secondary metabolites that tend to accumulate more in peel than in the pulp of the citrus fruit [9]. Flavonoids are a group of natural pigments with antifungal and anti-insect activities in plants and are thought to be involved in pathogen defense [10,11]. In addition, flavonoids have been reported to have anticancer and anti-allergy activities in the human body [12–14]. In citrus fruit, the contents of flavonoids decreased gradually during the ripening process. To date, however, the molecular mechanism that regulates flavonoid accumulation and degradation is still unclear [11].

Salicylic acid and jasmonic acid are two important plant hormones involved in plant development and growth, and their roles in regulating carotenoid and flavonoid accumulation have been studied in different plant species. In previous studies, it was reported that salicylic acid treatment increased flavonoid content in tea and decreased carotenoid content in plum [15,16]. In apple peel, it was found that methyl jasmonate treatment significantly promoted β-carotene accumulation [17]. In tomato, methyl jasmonate treatment promotes the lycopene accumulation through upregulating carotenoid biosynthetic pathway gene expression [18]. In citrus fruit, the research on the effects of salicylic acid and jasmonic acid is very limited. Huang et al. reported that salicylic acid treatment induced flavonoid and carotenoid accumulation and further enhanced antioxidant activity of preharvest navel oranges [19]. To date, however, the roles of salicylic acid and jasmonic acid in regulating the carotenoid and flavonoid accumulation in the juice sacs are still far from being elucidated in citrus fruit.

In the present study, to investigate how salicylic acid (SA) and methyl jasmonate (MeJA) treatments regulate flavonoid and carotenoid accumulation in the juice sacs of citrus fruit, we set up an in vitro system [20]. In the in vitro system, the juice sacs were cultured on MS medium supplemented with SA or MeJA, and the environmental conditions, such as light, water stress, and temperature were identical among the treatments. We evaluated the effects of plant hormones (abscisic acid and gibberellin), different colors of LED lights (blue and red), water stresses (sucrose and mannitol), and temperatures on carotenoid and ascorbate accumulation in the juice sacs of citrus fruit using this in vitro system [4,20–22]. In the present study, to elucidate the regulation of carotenoid and flavonoid metabolisms in response to SA and MeJA treatments in citrus fruit, the effects of SA and MeJA on carotenoid and flavonoid accumulation and the expression of genes related to carotenoid and flavonoid biosyntheses were investigated in the juice sacs in vitro. The results presented in this study provide new strategies to promote the accumulation of flavonoid and carotenoid and improve the quality of citrus fruit.

2. Materials and Methods

2.1. Plant Materials

Immature fruit of Satsuma mandarin (*Citrus unshiu* Marc.) were harvested from the Fujieda Farm of Shizuoka University (Shizuoka, Japan) during July 2016, and used as the plant materials.

2.2. In Vitro Culture System and Treatment

The fruits were surface sterilized by the following procedure: (1) rinsing the fruit in 70% ethanol, (2) soaking then in 1% (*w/v*) NaOCl for 1 h, and (3) rinsing the fruit with sterile water. After surface sterilization, juice sacs were excised and placed randomly on 10 mL of Murashige and Skoog (MS) medium supplemented with 1% (*w/v*) agar and 10% (*w/v*) sucrose [20] for two weeks at 20 °C in the dark. After that, the juice sacs were transplanted to a new MS medium supplemented with 1% (*w/v*) agar, 10% (*w/v*) sucrose, and 10 μM SA or MeJA, and then were cultured for 24, 48, and 72 h. Juice sacs cultured in the MS medium without plant hormones were used as a control. Finally, for each trial, the juice sacs were collected and immediately frozen in liquid nitrogen. The samples were kept at −80 °C until use.

2.3. Extraction and Determination of Flavonoids

The extraction and determination of flavonoids in citrus juice sacs were conducted using the methods described previously [23]. Flavonoids were extracted from the freeze-dried juice sacs using a methanol/dimethyl sulfoxide (50:50, *v/v*) extraction solution. The supernatant obtained by centrifugation (21,500× *g* for 10 min) was filtered through a 0.22 µm syringe filter (Shimadzu GLC, Tokyo, Japan), and the filtrate was analyzed by using a reverse-phase HPLC system (Jasco, Tokyo, Japan) fitted with a YMC-UltraHT pro C18 column (Waters, Milford, MA, USA). In this study, eriocitrin, narirutin, hesperidin, and poncirin were detected at 274 nm, and rhoifolin was detected at 338 nm (Supplementary Materials, Figure S1a,b). The content was estimated by the standard curves and expressed as mg g^{-1} of dry weight. The total flavonoids were calculated by summing all identified flavonoids. Flavonoid quantification was performed in three replicates.

2.4. Extraction and Determination of Carotenoids

The extraction and determination of carotenoids in citrus juice sacs were conducted using the methods described previously [24]. Pigments were extracted from the juice sacs using a hexane/acetone/ethanol (50:25:25, *v/v*) as extraction solvent containing 10% (*w/v*) magnesium carbonate basic. The organic solvents were completely evaporated under vacuum by rotary evaporator at maximum 35 °C. Afterwards, the extracts were saponified with 8 mL 20% (*w/v*) methanolic KOH and 12 mL diethyl ether containing 0.1% (*w/v*) 2,6-di-tert-butyl-4-methylphenol. Then, water-soluble extracts were removed by adding NaCl-saturated water. The pigments, repartitioned into the diethyl ether phase, were recovered and evaporated to dryness. Subsequently, the residue was redissolved in 5 mL of a tert-butyl-methyl-ether (TBME)/methanol (1:1, *v/v*) solution. After filtered through a 0.22 µm syringe filter (Shimadzu GLC, Tokyo, Japan), 1 mL of solvent were completely evaporated and redissolved in 100 µL TBME/methanol (1:1, *v/v*) solution. The samples were analyzed by using a reverse-phase HPLC system (Jasco, Tokyo, Japan) fitted with a YMC Carotenoid S-5 column (Waters, Milford, MA). In this study, all-*trans*-violaxanthin, 9-*cis*-violaxanthin, lutein, and β-cryptoxanthin, and β-carotene were detected at 452 nm (Supplementary Materials, Figure S1c). The contents of β-carotene, β-cryptoxanthin, lutein, all-*trans*-violaxanthin, and 9-*cis*-violaxanthin were estimated by the standard curves and expressed as µg g^{-1} of fresh weight [24]. The total carotenoids were calculated by summing all identified carotenoids. Carotenoid quantification was performed in three replicates.

2.5. Analysis of Gene Expression by Using Real-Time Quantitative RT-PCR

Total RNA was extracted from the juice sacs according to a previously reported method [25]. The total RNA was cleaned up with the RNeasy Mini kit (Qiagen, Hilden, Germany) with on-column DNase digestion. The reactions of reverse transcription (RT) were performed with 2 µg of purified RNA, a random hexamer, and TaqMan Reverse Transcription Reagents at 37 °C (Applied Biosystems, Foster City, CA, USA).

In Satsuma mandarin, the sequences of *CitUFGT2* and *CitWRKY70* have not been reported, thus, in the present study, we first amplified the cDNAs of *CitUFGT2* and *CitWRKY70* by PCR, and then the samples were sequenced by Fasmac Co., Ltd. TaqMan MGB probes and sets of primers for flavonoid biosynthetic genes (*CitCHS1*, *CitCHS2*, *CitCHI*, *CitUFGT2*, and *CitF3'H*), carotenoid biosynthetic genes (*CitPSY*, *CitPDS*, *CitLCYb1*, *CitLCYe*, *CitHYb*, and *CitNCED3*), and transcription factor gene (*CitWRKY70*) were designed with the Primer Express software (Supplementary Materials, Table S1). TaqMan real-time PCR was conducted with the TaqMan Universal PCR Master Mix (Applied Biosystems) using a StepOnePlus^{TM} Real-Time PCR System (Applied Biosystems). Each reaction mixture contained template cDNA, a TaqMan MGB Probe (250 nM), and primers (900 nM). The thermal cycling conditions consisted of 95 °C for 10 min, followed by 40 cycles of 95 °C for 15 s, and 60 °C for 1 min. Real-time quantitative RT-PCR was performed in three replicates for each sample.

2.6. RNA Isolation and Fluorescent Labeling of Probes

Total RNA was extracted by the methods of Ikoma et al. [25] from the juice sacs at 24, 48, and 72 h, after SA or methyl MeJA treatment or control. At least three independent RNA extractions were used in probe labeling for experimental reproducibility. The total RNA (400 ng) of all samples was labeled with the fluorescence Cy5, while the untreated 0 h sample was labeled with Cy3 according to the instructions for the Low RNA input linear amplification and labeling kit (Agilent Technologies, Palo Alto, CA, USA). Labeled cRNA was purified using the Qiagen RNeasy mini kit (Qiagen, Hilden, Germany). Hybridization and washing were performed according to the Agilent's instructions. Glass slides were hybridized overnight at 60 °C in a hybridization buffer containing a fragment of Cy3-or Cy5-labeled cRNA. After hybridization, slides were washed in 6 × SSC, 0.005% Triton X-100 for 10 min at room temperature and 0.1 × SSC, 0.005% Triton X-100 for 5 min at 4 °C. After drying the slides with gaseous nitrogen, hybridized slides were scanned with the use of an Agilent microarray scanner (Agilent Technologies, Palo Alto, CA, USA). The intensities of the Cy5 and Cy3 fluorescent signals from each spot were automatically normalized, and the ratio value (Cy5/Cy3) was calculated using Feature Extraction version 7.1 software (Linear & LOWESS analysis, Agilent). Data analysis was carried out using Subio (Subio).

2.7. Statistical Analysis

All values are shown as the mean ± SE. The data were analyzed, and Tukey's HSD test was used to analyze the differences amongst the treatments at $p < 0.05$ level.

3. Results

3.1. Effect of Salicylic Acid (SA) and Methyl Jasmonate (MeJA) on Flavonoid Content in Citrus Juice Sacs In Vitro

In this study, five flavonoids, eriocitrin, narirutin, hesperidin, poncirin, and rhoifolin were detected in juice sacs of Satsuma mandarin (Supplementary Materials, Figure S1a,b). In the juice sacs, hesperidin is the most predominantly accumulated, followed by narirutin. As compared with hesperidin and narirutin, the contents of eriocitrin, poncirin, and rhoifolin are relatively low in the juice sacs. To investigate how SA and MeJA treatments regulate flavonoid accumulation in the juice sacs of Satsuma mandarin, the changes in the contents of eriocitrin, narirutin, hesperidin, poncirin, and rhoifolin were determined in the juice sacs in vitro. As shown in Figure 1, significant changes in the flavonoid contents were observed after 72-h treatments of SA and MeJA. The content of rhoifolin was decreased, while the contents of eriocitrin, narirutin, and poncirin were significantly increased by the SA treatment, which were 0.6-fold, 0.3-fold, and 0.2-fold higher than that of the control, respectively. Therefore, the total flavonoid content after the SA treatment was higher than that of the control (111.8 vs. 99.2 mg·g^{-1}). Conversely, the contents of hesperidin, poncirin, and rhoifolin were significantly reduced after the MeJA treatment, and as a result the total flavonoid content was lower than that of the control.

3.2. Effect of SA and MeJA on Carotenoid Content in Citrus Juice Sacs In Vitro

The effects of SA and MeJA on the accumulation of the major carotenoids in the citrus juice sacs were also investigated. Although significant change in the total carotenoid content was not observed, the accumulation of individual carotenoid, such as β-cryptoxanthin and all-*trans*-violaxanthin, was affected by the SA treatment (Figure 2). Indeed, the content of all-*trans*-violaxanthin decreased, while that of β-cryptoxanthin, which accounts for about 30% of total carotenoid, increased after the SA treatment. On the contrary, the total carotenoid content was much lower than the control after the MeJA treatment. In particular, only the content of β-cryptoxanthin increased up to 0.24-fold with respect to the control; whilst the concentration of all-*trans*-violaxanthin, 9-*cis*-violaxanthin, and lutein significantly decreased (Figure 2).

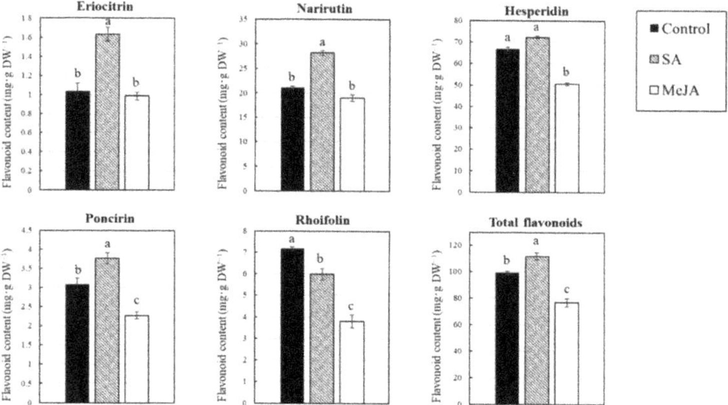

Figure 1. Effect of salicylic acid (SA) and methyl jasmonate (MeJA) treatments on the flavonoid content in the juice sacs of Satsuma mandarin in vitro. The total flavonoid was calculated by summing eriocitrin, narirutin, hesperidin, poncirin, and rhoifolin. Columns and bars represent the means and SE (n = 3), respectively. Different letters indicate significant differences at the 5% level by Tukey's HSD test.

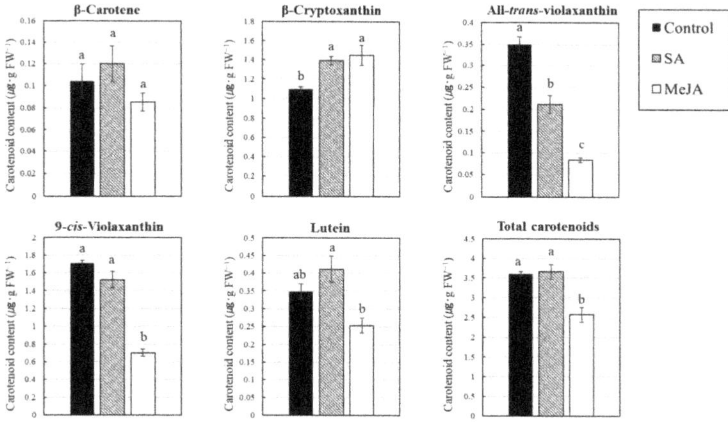

Figure 2. Effect of salicylic acid (SA) and methyl jasmonate (MeJA) treatments on the carotenoid content in the juice sacs of Satsuma mandarin in vitro. The total carotenoid was calculated by summing β-carotene, β-cryptoxanthin, all-*trans*-violaxanthin, 9-*cis*-violaxanthin, and lutein. Columns and bars represent the means and SE (n = 3), respectively. Different letters indicate significant differences at the 5% level by Tukey's HSD test.

3.3. Effect of SA and MeJA on Expression of the Genes Related to Flavonoid and Carotenoid Accumulation in Citrus Juice Sacs

In this study, the changes in the expression of the genes related to flavonoid and carotenoid metabolisms in response to SA and MeJA treatments were investigated by using real-time PCR. The results showed that the expression of *CitCHS1* was downregulated, while the expression of *CitCHS2, CitCHI, CitUFGT2,* and *CitF3′H* was significantly upregulated by the SA treatment (Figure 3). In the MeJA treatment, the expression of *CitCHS1* and *CitF3′H* was downregulated, while the expression of *CitCHI* and *CitUFGT2* was upregulated in the citrus juice sacs (Figure 3). Furthermore, the expression of five carotenoid biosynthetic genes (*CitPSY, CitPDS CitLCYe, CitLCYb1* and *CitHYb*) and one carotenoid catabolic gene (*CitNCED3*) was investigated in the citrus juice sacs. In the SA treatment, the expression of *CitPSY*, *CitPDS*, and *CitHYb* was upregulated, while that of *CitNCED3* was downregulated. Instead, in the MeJA treatment, the expression of *CitHYb* was downregulated, and that of the other carotenoid metabolic genes was not significantly affected (Figure 4).

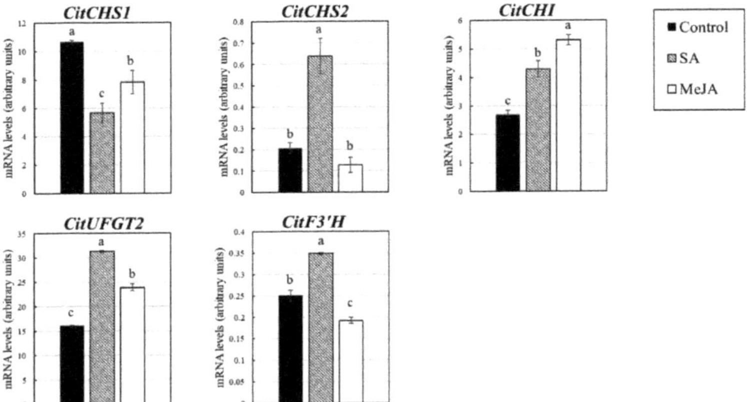

Figure 3. Effect of salicylic acid (SA) and methyl jasmonate (MeJA) treatments on the expression of flavonoid metabolism related genes in the juice sacs of Satsuma mandarin in vitro. The mRNA levels were analyzed by TaqMan real-time RT-PCR, and the expression of 18S ribosomal RNA (rRNA) was used as a control to normalize the raw data. Columns and bars represent the means and SE ($n = 3$), respectively. Different letters indicate significant differences at the 5% level by Tukey's HSD test. CHS1, chalcone synthase1; CHS2, chalcone synthase2; CHI, chalcone isomerase; UFGT2, flavanone 7-O-glucosyltransferase2; F3'H, flavonoid-3'-hydroxylase.

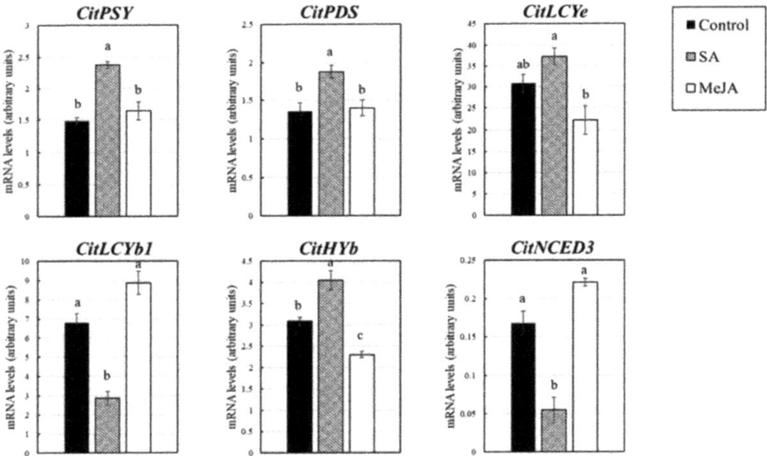

Figure 4. Effect of salicylic acid (SA) and methyl jasmonate (MeJA) treatments on the expression of carotenoid metabolism related genes in the juice sacs of Satsuma mandarin in vitro. The mRNA levels were analyzed by TaqMan real-time RT-PCR, and the expression of 18S ribosomal RNA (rRNA) was used as a control to normalize the raw data. Columns and bars represent the means and SE ($n = 3$), respectively. Different letters indicate significant differences at the 5% level by Tukey's HSD test. PSY, phytoene synthase; PDS, 15-*cis*-phytoene desaturase; LCYe, lycopene epsilon-cyclase; LCYb1, lycopene beta-cyclase1; HYb, beta-carotene 3-hydroxylase; NCED3, 9-*cis*-epoxycarotenoid dioxygenase3.

3.4. Identification of Transcription Factor in Response to Salicylic Acid and Methyl Jasmonate Treatments

In the present study, differentially expressed genes were investigated in the SA and MeJA treatments by using microarray analysis. As compared with the control, the expressions of 1616 genes were found to be significantly changed after 24-, 48-, and 72-h treatments of SA (Figure 5a and Supplementary Materials, Table S2). In particular, the expression of 1286 genes was upregulated, while the expression of 330 genes was downregulated by the SA treatment. In the MeJA treatment, there were 4914 genes that were differentially expressed as compared with the control (Figure 5b and Supplementary Materials, Table S3). Among them, the expression of 4404 genes was upregulated, while the expression of 510 genes was downregulated by the MeJA treatment. In this study, we found that SA treatment upregulated the key genes related to flavonoid and carotenoid biosynthesis (e.g., *CitCHS2*, *CitF3'H*, and *CitHYb*) in the juice sacs in vitro. In contrast, the expression of these key genes was downregulated by the MeJA treatment in the juice sacs in vitro. To identify the transcriptional factors that might be involved in the regulation of flavonoid and carotenoid accumulation, we further screened the genes which were upregulated in the SA treatment while downregulated in the MeJA. As shown in Figure 6, we found that there were three genes that were upregulated by SA treatment while downregulated by the MeJA treatment (Cicev10012055m, Ciclev10022863m, and Ciclev10025958m; Table 1). Among them, a WRKY DNA-binding protein 70 was identified (Cicev10012055m), which exhibited 45.6% homology to Arabidopsis *WRKY70*. Expression analysis of *CitWRKY70* by using real-time PCR confirmed that it was a 9.4-fold upregulated by the SA treatment, and a 14.4-fold downregulated by the MeJA treatment as compared with the control (Figure 7).

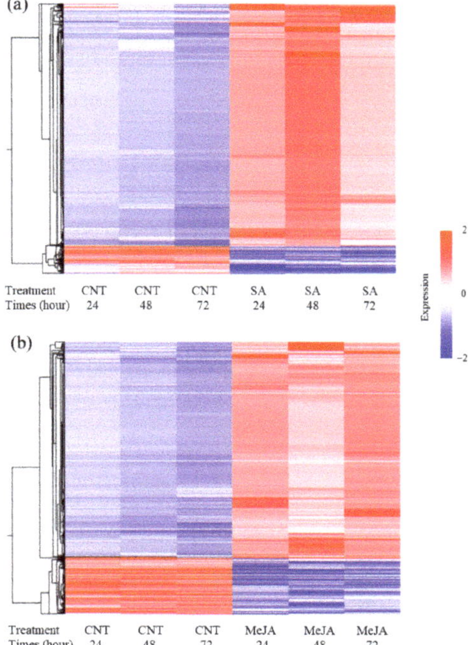

Figure 5. Hierarchical cluster analysis of 1616 salicylic acid-responsive genes and 4914 methyl jasmonate-responsive genes with more than 2-fold expression changes between salicylic acid or methyl jasmonate and control treatments. (**a**) Salicylic acid (SA)/control (CNT) signal intensity ratio; (**b**) Methyl jasmonate (MeJA)/control (CNT) signal intensity ratio. The color scale indicates a signal intensity of each gene. Tree at the left side of the matrix represents gene relationship.

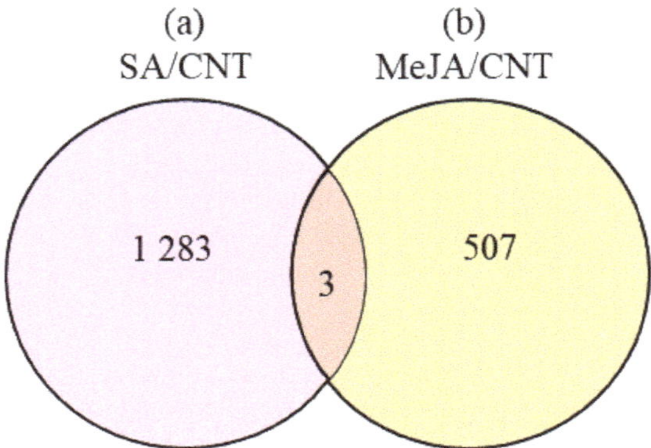

Figure 6. Two-set Venn diagram analysis of salicylic acid (SA) and methyl jasmonate (MeJA)-responsive genes in the juice sacs in vitro. (**a**) A set of 1286 salicylic acid-responsive genes whose expression is more than 2-fold increased by salicylic acid treatment; (**b**) A set of 510 methyl jasmonate-responsive genes whose expression is more than 2-fold decreased by methyl jasmonate treatment. CNT, control.

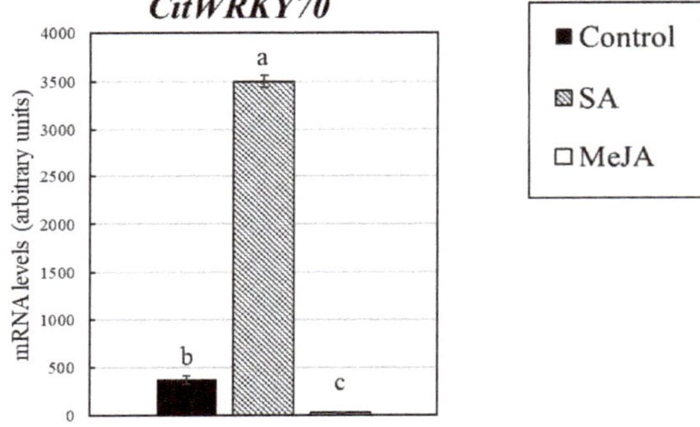

Figure 7. Effects of salicylic acid (SA) and methyl jasmonate (MeJA) treatments on the expression of *CitWRKY70* in the juice sacs of Satsuma mandarin in vitro. The mRNA levels were analyzed by TaqMan real-time RT-PCR, and the expression of 18S ribosomal RNA (rRNA) was used as a control to normalize the raw data. Columns and bars represent the means and SE ($n = 3$), respectively. Different letters indicate significant differences at the 5% level by Tukey's HSD test. *WRKY70*, WRKY DNA-binding protein 70.

Table 1. Details of three genes that were more than 2-fold upregulated in salicylic acid (SA) treatment and 2-fold downregulated in methyl jasmonate (MeJA) treatment. CNT, control.

ID	mRNA	JGI_AGI	Abbreviations	Enzyme	SA/CNT			MeJA/CNT		
					24 h	48 h	72 h	24 h	48 h	72 h
C22820	Ciclev10022863m			PROTEIN NIM1-INTERACTING 1	2.53	2.26	4.18	0.31	0.42	0.35
C25898	Ciclev10025958m	AT4G10490.1	DMR-6	2-oxoglutarate (2OG) and Fe (II)-dependent	17.93	12.21	67.54	0.23	0.12	0.10
C12031	Ciclev10012055m	AT3G56400.1	WRKY70	WRKY DNA-binding protein 70	6.52	5.51	10.30	0.30	0.16	0.28

4. Discussion

4.1. Effect of SA on the Flavonoid and Carotenoid Accumulation in Citrus Juice Sacs In Vitro

Flavonoids and carotenoids are two important secondary metabolites that accumulate in plants. To date, the role of SA in flavonoid accumulation has been investigated in different plant species such as tea and cucumber [15,26]. SA and flavonoids are both phenylpropanoid derivatives, and recent studies have suggested the SA treatment affected the flavonoid biosynthesis in plants [27]. In tea, it was reported that exogenous methyl salicylate treatment increased the expression of flavonoid biosynthesis-related genes, such as *CHS* and *CHI*, and as a result the total content of flavonoid was enhanced by the methyl salicylate treatment [15]. In cucumber, 0.5 mM salicylic acid treatment was found to be effective to enhance the contents of total phenolic compound and total flavonoid [26]. In the present study, the results showed that SA treatment induced the accumulation of eriocitrin, narirutin, and poncirin in the citrus juice sacs in vitro (Figure 1). The gene expression analysis showed that the expression of *CitCHS2*, *CitUFGT2*, and *CitF3′H* was markedly increased by the SA treatment (Figure 3). As previously reported, *CitCHS2*, which was highly expressed in the young fruit, was an important gene regulating the biosynthesis of flavonoid in citrus fruit [28]. In the present study, the high expression of *CitCHS2* could lead to enhance the contents of eriocitrin, narirutin, hesperidin, and poncirin in the SA-treated juice sacs. In addition, CitF3′H, which catalyzes the hydroxylation of β-rings at the 3′-position, is a rate limiting enzyme for flavonoid biosynthesis in plant [29]. In this study, the high expression of *CitF3′H* in the SA treatment could contribute to increasing the content of eriocitrin, which has a hydroxy group on the β-rings at the 3′-position, in the citrus juice sacs.

Different from flavonoid, the roles of SA in carotenoid accumulation were controversial in plants. In wheat and moong seedings, total carotenoid content, size of xanthophyll pool, and de-epoxidation rate were induced along with an increase in SA concentration [30]. In plum, in contrast, the preharvest treatment of salicylate significantly reduced the total carotenoid content [16]. In the present study, the results showed that the total carotenoid content was not significantly affected by the SA treatment, but the content of β-cryptoxanthin, which was the major carotenoid accumulated in the juice sacs, was clearly enhanced by the SA treatment. The gene expression results suggested that the higher expression of carotenoid biosynthetic genes, *CitPSY*, *CitPDS*, and *CitHYb* and lower expression of carotenoid catabolic gene *CitNCED3* could lead to enhance β-cryptoxanthin content in the juice sacs treated by SA.

Summarizing, we found that SA treatment was effective to enhance the contents of eriocitrin, narirutin, and poncirin and β-cryptoxanthin in the citrus juice sacs of Satsuma mandarin. Overall, our findings were consistent with a previous study, in which the preharvest treatment of salicylate caused an increase of carotenoids and flavonoids contents in navel oranges [19]. Flavonoids and carotenoids are important antioxidant compounds in citrus fruits with beneficial effects to human health [31]. Thus, the results presented in this study suggested that SA treatment could be an effective method to improve the quality and nutritional values of citrus fruit.

4.2. Effect of MeJA on the Flavonoid and Carotenoid Accumulation in Citrus Juice Sacs In Vitro

Jasmonic acid is a plant hormone that acts as a signal of plant responses to biotic and abiotic stresses, including the regulation of the bioactive compounds accumulation in plants [32]. It has been reported that MeJA was effective to induce flavonoid accumulation in apples [33], red Chinese pear [34], red raspberry [35], and blackberry [36]. To date, the studies of MeJA-regulated flavonoid accumulation have been mainly focused on anthocyanin and phenolic biosyntheses, while its effects on the flavonoids that are specifically accumulated in the citrus fruit have been largely unknown. In the present study, the results showed that the MeJA treatment significantly decreased the contents of hesperidin, poncirin, and rhoifolin in the juice sacs of citrus fruit, and as a result the total flavonoid content was much lower than that of the control. The gene expression results showed the downregulation of expression of *CitCHS1*, *CitCHS2*, and *CitF3′H* was consistent with the decrease in the flavonoids contents in the juice sacs. In addition, we found that the MeJA treatment not only inhibited flavonoid accumulation, but also caused a significant decrease of carotenoids in the juice sacs. Indeed, the contents of all-*trans*-violaxanthin, 9-*cis*-violaxanthin, lutein, as well as total carotenoid were much lower than that of the control. Moreover, the expression of *CitHYb*, which is a key gene responsible for xanthophyll biosynthesis, was downregulated by the MeJA treatment. Recently, the roles of jasmonate on carotenoid accumulation has been reported in lettuce, tomato, and maize kernels, in which it was indicated that jasmonate positively regulated the accumulation of lycopene and lutein, and a deficiency of jasmonate led to a great reduction of lycopene in tomato fruit [18,37,38]. However, the effect of jasmonate on carotenoids accumulation was dose dependent, i.e., when the concentration of MeJA was higher than 0.5 µM, the carotenoid biosynthesis was inhibited in tomato and maize kernels. In the present study, we treated the juice sacs with 10 µM MeJA and found that the carotenoids content was significantly decreased in the juice sacs. These results were in agreement with previous studies in tomato and maize kernels [38]. In future studies, research on the effects of different concentrations of MeJA on carotenoid biosynthesis is expected, which should contribute to improving the carotenoid accumulation in the juice sacs of citrus fruit.

4.3. Identification of Transcription Factor in Response to SA and MeJA Treatments

WRKY70 is a unique member of the WRKY transcriptional factor family controlling plant defense, senescence, and developmental processes [39–43]. Li et al. [44] reported that *WRKY70* was a common component in salicylic acid and jasmonic acid mediated signal pathways, and played a pivotal role in determining the balance between salicylic acid-dependent and jasmonic acid-dependent defense pathway. In Arabidopsis, it was found that the expression of *WRKY70* was activated by SA and repressed by MeJA. In citrus, however, the research on *WRKY70* has not been reported. In the present study, differentially expressed genes in the SA and MeJA treatments were investigated by using microarray analysis. We found that the expression of *CitWRKY70* was induced by the SA treatment, while suppressed by the MeJA treatment in the juice sacs in vitro. In our study, it was showed that the expression of key genes related to flavonoid and carotenoid biosyntheses (e.g., *CitCHS2*, *CitF3′H*, and *CitHYb*) was also induced by SA treatment, while suppressed by the MeJA treatment. Thus, it was deduced that *CitWRKY70* could be involved in the regulation of flavonoid and carotenoid accumulation in response to the treatments with the two phytohormones. In the future, more in-depth researches on the function of *WRKY70* in flavonoid and carotenoid biosyntheses are still needed, which would contribute to elucidating the regulation mechanisms of flavonoid and carotenoid accumulation in response to SA and MeJA treatments in citrus fruit.

5. Conclusions

In the present study, the effects of SA and MeJA on flavonoid and carotenoid accumulation were investigated in the juice sacs of Satsuma mandarin in vitro. The results suggested that the plant hormones SA and MeJA are important factors regulating flavonoid and carotenoid accumulation in the citrus juice sacs. The SA treatment was effective to enhance the contents of eriocitrin, narirutin, poncirin, and β-cryptoxanthin in the citrus juice sacs. Whereas the carotenoid and flavonoid accumulation was inhibited by the MeJA treatment in the citrus juice sacs in vitro. The gene expression results suggested that the changes in the flavonoid and carotenoid accumulation in the SA and MeJA treatments were highly regulated at the transcriptional level. In addition, a transcriptional factor *CitWRKY70* was identified in this study by using the microarray analysis. The changes in the expression of *CitWRKY70* was consistent with those of the key flavonoid and carotenoid biosynthetic genes, indicating that *CitWRKY70* could be involved in the regulation of flavonoid and carotenoid accumulation in response to SA and MeJA treatments. In this study, the regulation of flavonoid and carotenoid accumulation by SA and MeJA treatments were investigated, and their molecular mechanisms were deeply discussed in citrus fruit. The results present herein contribute to elucidating the roles of SA and MeJA in plants and provide new strategies to modify flavonoid and carotenoid accumulation in citrus fruit. Moreover, as the flavonoid and carotenoid contents were significantly enhanced by the SA treatment in this study, it was indicated that application of SA could be an effective method to improve the nutritional and commotional values of citrus fruit. In the future study, more in-depth researches on the function of *WRKY70* and effects of different concentrations of MeJA in citrus fruit are expected, which would contribute to further elucidating the regulation mechanisms of flavonoid and carotenoid accumulation in response to SA and MeJA treatments in citrus fruit.

Supplementary Materials: The following are available online at http://www.mdpi.com/2076-3417/10/24/8916/s1, Table S1: Primer sequences and TaqMan MGB Probes used for the TaqMan real-time RT-PCR of the genes related to plant hormones, flavonoid and carotenoid metabolisms., Table S2: Representative salicylic acid responsive genes showing differential expression (>2-fold) between salicylic acid and control treatments (salicylic acid/ control ratio). Table S3: Representative methyl jasmonate responsive genes showing differential expression (>2-fold) between methyl jasmonate and control treatments (methyl jasmonate/control ratio). Figure S1: Chromatograms in the flavonoid and carotenoid in the juice sacs of citrus fruit in vitro.

Author Contributions: Conceptualization, G.M. and M.K.; methodology, T.S., H.F., and T.E.; validation, R.Y., G.M., L.Z., and M.K.; formal analysis, R.Y. and G.M.; investigation, R.Y., M.H., and G.M.; resources, M.Y. and K.Y.; data curation, R.Y., G.M., L.Z., and M.K.; writing—original draft preparation, R.Y. and G.M.; writing—review and editing, L.Z. and M.K.; visualization R.Y., G.M., and L.Z.; supervision, M.K.; project administration, G.M. and M.K.; funding acquisition, G.M. and M.K. All authors have read and agreed to the published version of the manuscript.

Funding: This work was supported by KAKENHI grant numbers JP26292015 (to M.K.) and JP19K06030 (to G.M.) from Japan Society for the Promotion of Science (JSPS).

Conflicts of Interest: The authors declare no conflict of interest.

References

1. Fujii, H.; Shimada, T.; Sugiyama, A.; Nishikawa, F.; Endo, T.; Nakano, M.; Ikoma, Y.; Shimizu, T.; Omura, M. Profiling ethylene-responsive genes in mature mandarin fruit using a citrus 22K oligoarray. *Plant Sci.* **2007**, *173*, 340–348. [CrossRef]

2. Giovannoni, J.J. Genetic Regulation of Fruit Development and Ripening. *Plant Cell* **2004**, *16* (Suppl. 1), S170–S180. [CrossRef]

3. Takayanagi, K.; Morimoto, S.; Shirakura, Y.; Mukai, K.; Sugiyama, T.; Tokuji, Y.; Ohnishi, M. Mechanism of Visceral Fat Reduction in Tsumura Suzuki Obese, Diabetes (TSOD) Mice Orally Administered β-Cryptoxanthin from Satsuma Mandarin Oranges (Citrus Unshiu Marc). *J. Agric. Food Chem.* **2011**, *59*, 12342–12351. [CrossRef]

4. Ma, G.; Zhang, L.; Iida, K.; Madono, Y.; Yungyuen, W.; Yahata, M.; Yamawaki, K.; Kato, M. Identification and quantitative analysis of β-cryptoxanthin and β-citraurin esters in satsuma mandarin fruit during the ripening process. *Food Chem.* **2017**, *234*, 356–364. [CrossRef]

5. Yamaguchi, M. Role of carotenoid β-cryptoxanthin in bone homeostasis. *J. Biomed. Sci.* **2012**, *19*, 1–13. [CrossRef] [PubMed]

6. Iskandar, A.R.; Liu, C.; Smith, D.E.; Hu, K.-Q.; Choi, S.-W.; Ausman, L.M.; Wang, X.-D. β-Cryptoxanthin Restores Nicotine-Reduced Lung SIRT1 to Normal Levels and Inhibits Nicotine-Promoted Lung Tumorigenesis and Emphysema in A/J Mice. *Cancer Prev. Res.* **2013**, *6*, 309–320. [CrossRef] [PubMed]

7. Sugiura, M.; Nakamura, M.; Ogawa, K.; Ikoma, Y.; Ando, F.; Shimokata, H.; Yano, M. Dietary patterns of antioxidant vitamin and carotenoid intake associated with bone mineral density: Findings from Post-Menopausal Japanese Female Subjects. *Osteoporos. Int.* **2011**, *22*, 143–152. [CrossRef] [PubMed]

8. Ma, G.; Zhang, L.; Sugiura, M.; Kato, M. Citrus and health. In *The Genus Citrus*; Elsevier: Amsterdam, The Netherlands, 2020; pp. 495–511.

9. Nogata, Y.; Sakamoto, K.; Shiratsuchi, H.; Ishii, T.; YANO, M.; Ohta, H. Flavonoid Composition of Fruit Tissues of Citrus Species. *Biosci. Biotechnol. Biochem.* **2006**, *70*, 178–192. [CrossRef] [PubMed]

10. Kim, B.-G.; Kim, H.; Hur, H.-G.; Lim, Y.; Ahn, J.-H. Regioselectivity of 7-O-methyltransferase of poplar to flavones. *J. Bbiotechnol.* **2006**, *126*, 241–247. [CrossRef]

11. Napal, G.N.D.; Carpinella, M.C.; Palacios, S.M. Antifeedant activity of ethanolic extract from Flourensia oolepis and isolation of pinocembrin as its active principle compound. *Bioresour. Technol.* **2009**, *100*, 3669–3673. [CrossRef]

12. Cheng, H.-L.; Hsieh, M.-J.; Yang, J.-S.; Lin, C.-W.; Lue, K.-H.; Lu, K.-H.; Yang, S.-F. Nobiletin inhibits human osteosarcoma cells metastasis by blocking ERK and JNK-mediated MMPs expression. *Oncotarget* **2016**, *7*, 35208. [CrossRef]

13. Goh, J.X.H.; Tan, L.T.-H.; Goh, J.K.; Chan, K.G.; Pusparajah, P.; Lee, L.-H.; Goh, B.-H. Nobiletin and Derivatives: Functional Compounds from Citrus Fruit Peel for Colon Cancer Chemoprevention. *Cancers* **2019**, *11*, 867. [CrossRef] [PubMed]

14. Takano, H.; Osakabe, N.; Sanbongi, C.; Yanagisawa, R.; Inoue, K.-I.; Yasuda, A.; Natsume, M.; Baba, S.; Ichiishi, E.-I.; Yoshikawa, T. Extract of Perilla frutescens Enriched for Rosmarinic Acid, a Polyphenolic Phytochemical, Inhibits Seasonal Allergic Rhinoconjunctivitis in Humans. *Exp. Biol. Med.* **2004**, *229*, 247–254. [CrossRef] [PubMed]

15. Li, X.; Zhang, L.-P.; Zhang, L.; Yan, P.; Ahammed, G.J.; Han, W.-Y. Methyl Salicylate Enhances Flavonoid Biosynthesis in Tea Leaves by Stimulating the Phenylpropanoid Pathway. *Molecules* **2019**, *24*, 362. [CrossRef] [PubMed]

16. Martínez-Esplá, A.; Zapata, P.J.; Valero, D.; Martínez-Romero, D.; Díaz-Mula, H.M.; Serrano, M. Preharvest treatments with salicylates enhance nutrient and antioxidant compounds in plum at harvest and after storage. *J. Sci. Food Agric.* **2018**, *98*, 2742–2750. [CrossRef] [PubMed]

17. Pérez, A.G.; Sanz, C.; Richardson, D.G.; Olías, J.M. Methyl Jasmonate Vapor Promotes β-Carotene Synthesis and Chlorophyll Degradation in Golden Delicious Apple Peel. *J. Plant Growth Regul.* **1993**, *12*, 163. [CrossRef]

18. Liu, L.; Wei, J.; Zhang, M.; Zhang, L.; Li, C.; Wang, Q. Ethylene independent induction of lycopene biosynthesis in tomato fruits by jasmonates. *J. Exp. Bot.* **2012**, *63*, 5751–5761. [CrossRef]

19. Huang, R.; Xia, R.; Lu, Y.; Hu, L.; Xu, Y. Effect of pre-harvest salicylic acid spray treatment on post-harvest antioxidant in the pulp and peel of 'Cara cara' navel orange (*Citrus sinenisis* L. Osbeck). *J. Sci. Food Agric.* **2008**, *88*, 229–236. [CrossRef]

20. Zhang, L.; Ma, G.; Kato, M.; Yamawaki, K.; Takagi, T.; Kiriiwa, Y.; Ikoma, Y.; Matsumoto, H.; Yoshioka, T.; Nesumi, H. Regulation of carotenoid accumulation and the expression of carotenoid metabolic genes in citrus juice sacs in vitro. *J. Exp. Bot.* **2012**, *63*, 871–886. [CrossRef]

21. Ma, G.; Zhang, L.; Yungyuen, W.; Sato, Y.; Furuya, T.; Yahata, M.; Yamawaki, K.; Kato, M. Accumulation of carotenoids in a novel citrus cultivar 'Seinannohikari' during the fruit maturation. *Plant Physiol. Biochem.* **2018**, *129*, 349–356. [CrossRef]

22. Zhang, L.; Ma, G.; Yamawaki, K.; Ikoma, Y.; Matsumoto, H.; Yoshioka, T.; Ohta, S.; Kato, M. Effect of Blue LED light intensity on carotenoid accumulation in citrus juice sacs. *Physiol. Plant.* **2015**, *188*, 58–63. [CrossRef] [PubMed]

23. Seoka, M.; Ma, G.; Zhang, L.; Yahata, M.; Yamawaki, K.; Kan, T.; Kato, M. Expression and functional analysis of the nobiletin biosynthesis-related gene *CitOMT* in citrus fruit. *Sci. Rep.* **2020**, *10*, 15288. [CrossRef] [PubMed]

24. Kato, M.; Ikoma, Y.; Matsumoto, H.; Sugiura, M.; Hyodo, H.; Yano, M. Accumulation of Carotenoids and Expression of Carotenoid Biosynthetic Genes during Maturation in Citrus Fruit. *Physiol. Plant.* **2004**, *134*, 824–837. [CrossRef] [PubMed]

25. Ikoma, Y.; Yano, M.; Ogawa, K.; Yoshioka, T.; Xu, Z.C.; Hisada, S.; Omura, M.; Moriguchi, T. Isolation and Evaluation of RNA from Polysaccharide-Rich Tissues in Fruit for Quality by CDNA Library Construction and RT-PCR. *J. Jpn. Soc. Hort. Sci.* **1996**, *64*, 809–814. [CrossRef]

26. Preciado-Rangel, P.; Reyes-Pérez, J.J.; Ramírez-Rodríguez, S.C.; Salas-Pérez, L.; Fortis-Hernández, M.; Murillo-Amador, B.; Troyo-Diéguez, E. Foliar Aspersion of Salicylic Acid Improves Phenolic and Flavonoid Compounds, and Also the Fruit Yield in Cucumber (Cucumis Sativus L.). *Plants* **2019**, *8*, 44. [CrossRef]

27. Gondor, O.K.; Janda, T.; Soós, V.; Pál, M.; Majláth, I.; Adak, M.K.; Balázs, E.; Szalai, G. Salicylic Acid Induction of Flavonoid Biosynthesis Pathways in Wheat Varies by Treatment. *Front. Plant Sci.* **2016**, *7*, 1447. [CrossRef]

28. Moriguchi, T.; Kita, M.; Tomono, Y.; Endo-Inagaki, T.; Omura, M. Gene expression in flavonoid biosynthesis: Correlation with flavonoid accumulation in developing citrus fruit. *Physiol Plant.* **2001**, *111*, 66–74. [CrossRef]

29. Baba, S.A.; Ashraf, N. Functional characterization of flavonoid 3'-hydroxylase, CsF3' H, from *Crocus sativus* L: Insights into substrate specificity and role in abiotic stress. *Arch. Biochem. Biophys.* **2019**, *667*, 70–78. [CrossRef]

30. Moharekar, S.T.; Lokhande, S.D.; Hara, T.; Tanaka, R.; Tanaka, A.; Chavan, P.D. Effect of salicylic acid on chlorophyll and carotenoid contents of wheat and moong seedlings. *Photosynthetica* **2003**, *41*, 315. [CrossRef]

31. Giménez, M.J.; Valverde, J.M.; Valero, D.; Zapata, P.J.; Castillo, S.; Serrano, M. Postharvest methyl salicylate treatments delay ripening and maintain quality attributes and antioxidant compounds of 'Early Lory' sweet cherry. *Postharvest Biol. Technol.* **2016**, *117*, 102–109. [CrossRef]

32. Yang, Y.-X.; Ahammed, G.J.; Wu, C.; Fan, S.; Zhou, Y.-H. Crosstalk among Jasmonate, Salicylate and Ethylene Signaling Pathways in Plant Disease and Immune Responses. *Curr. Protein Pept. Sci.* **2015**, *16*, 450–461. [CrossRef] [PubMed]

33. Rudell, D.R.; Mattheis, J.P.; Fan, X.; Fellman, J.K. Methyl Jasmonate Enhances Anthocyanin Accumulation and Modifies Production of Phenolics and Pigments InFuji'Apples. *J. Am. Soc. Hort. Sci.* **2002**, *127*, 435–441. [CrossRef]

34. Ni, J.; Zhao, Y.; Tao, R.; Yin, L.; Gao, L.; Strid, A.; Qian, M.; Li, J.; Li, Y.; Shen, J. Ethylene mediates the branching of the jasmonate-induced flavonoid biosynthesis pathway by suppressing anthocyanin biosynthesis in red Chinese pear fruits. *Plant Biotechnol. J.* **2020**, *18*, 1223–1240. [CrossRef] [PubMed]

35. Flores, G.; del Castillo, M.L.R. Influence of preharvest and postharvest methyl jasmonate treatments on flavonoid content and metabolomic enzymes in red raspberry. *Postharvest Biol. Technol.* **2014**, *97*, 77–82. [CrossRef]

36. Wang, S.Y.; Bowman, L.; Ding, M. Methyl jasmonate enhances antioxidant activity and flavonoid content in blackberries (*Rubus* sp.) and promotes antiproliferation of human cancer cells. *Food Chem.* **2008**, *107*, 1261–1269. [CrossRef]

37. Yuan, H.; Zhang, J.; Nageswaran, D.; Li, L. Carotenoid metabolism and regulation in horticultural crops. *Hortic. Res.* **2015**, *2*, 1–11. [CrossRef]

38. Luo, H.; He, W.; Li, D.; Bao, Y.; Riaz, A.; Xiao, Y.; Song, J.; Liu, C. Effect of methyl jasmonate on carotenoids biosynthesis in germinated maize kernels. *Food Chem.* **2020**, *307*, 125525. [CrossRef]

39. Chen, C.; Chen, Z. Potentiation of Developmentally Regulated Plant Defense Response by AtWRKY18, a Pathogen-Induced Arabidopsis Transcription Factor. *Physiol. Plant* **2002**, *129*, 706–716. [CrossRef]

40. Du, L.; Chen, Z. Identification of genes encoding receptor-like protein kinases as possible targets of pathogen-and salicylic acid-induced WRKY DNA-binding proteins in *Arabidopsis*. *Plant J.* **2000**, *24*, 837–847. [CrossRef]

41. Robatzek, S.; Somssich, I.E. A new member of the *Arabidopsis* WRKY transcription factor family, *At*WRKY6, is associated with both senescence-and defence-related processes. *Plant J.* **2001**, *28*, 123–133. [CrossRef]

42. Xie, T.; Chen, C.; Li, C.; Liu, J.; Liu, C.; He, Y. Genome-wide investigation of WRKY gene family in pineapple: Evolution and expression profiles during development and stress. *BMC Genom.* **2018**, *19*, 490. [CrossRef] [PubMed]

43. Yu, D.; Chen, C.; Chen, Z. Evidence for an Important Role of WRKY DNA Binding Proteins in the Regulation of *NPR1* Gene Expression. *Plant Cell* **2001**, *13*, 1527–1540. [CrossRef] [PubMed]

44. Li, J.; Brader, G.; Palva, E.T. The WRKY70 Transcription Factor: A Node of Convergence for Jasmonate-Mediated and Salicylate-Mediated Signals in Plant Defense. *Plant Cell* **2004**, *16*, 319–331. [CrossRef] [PubMed]

Publisher's Note: MDPI stays neutral with regard to jurisdictional claims in published maps and institutional affiliations.

© 2020 by the authors. Licensee MDPI, Basel, Switzerland. This article is an open access article distributed under the terms and conditions of the Creative Commons Attribution (CC BY) license (http://creativecommons.org/licenses/by/4.0/).

 *applied sciences*

Article

Modelling of Carotenoids Content in Red Clover Sprouts Using Light of Different Wavelength and Pulsed Electric Field

Ilona Gałązka-Czarnecka [1,*], Ewa Korzeniewska [2,*], Andrzej Czarnecki [1], Paweł Kiełbasa [3] and Tomasz Dróżdż [3]

[1] Faculty of Biotechnology and Food Sciences, Lodz University of Technology, Stefanowskiego 4/10, 90-924 Łódź, Poland; andrzej.czarnecki@p.lodz.pl

[2] Institute of Electrical Engineering Systems, Lodz University of Technology, Stefanowskiego 18/22, 90-924 Łódź, Poland

[3] Faculty of Production and Power Engineering, University of Agriculture in Krakow, ul. Balicka 116 B, 30-149 Kraków, Poland; pawel.kielbasa@urk.edu.pl (P.K.); tomasz.drozdz@office.urk.edu.pl (T.D.)

[*] Correspondence: ilona.czarnecka@p.lodz.pl (I.G.-C.); ewa.korzeniewska@p.lodz.pl (E.K.); Tel.: +48-42-631-34-63 (I.G.-C.)

Received: 4 May 2020; Accepted: 11 June 2020; Published: 16 June 2020

Abstract: The paper presents the results of influence the light of different wavelengths and pulsed electric fields on the content of carotenoids. Seeds germination was carried out in a climatic chamber with phytotron system. The experiment was carried out under seven growing conditions differing in light-emitting diode (LED) wavelengths and using pulsed electric fields (PEFs) with different strength applied before sowing. Cultivation of the sprouts was carried out for seven days at relative humidity 80% and 20 ± 1 °C. Different light wavelengths were used during cultivation: white light (380–780 nm), UVA (340 nm), blue (440 nm), and red (630 nm). In addition, the pulsed electric field (PEF) with three values of strength equal to 1, 2.5 and 5 kV/cm, respectively, was applied to three series of sprouts before sowing. Sprouts treated with the PEF were grown under white light (380–780 nm). The light exposure time for all experimental series of sprouts was 12/12 h (12 h light, 12 h dark for seven days). Lutein is the dominant carotenoid in germinating red clover seeds, the content of which varies from 743 mg/kg in sprouts grown in red light, 862 mg/kg in sprouts grown in UVA, to 888 mg/kg in sprouts grown in blue light. Blue light in the cultivation of red clover sprouts had the most beneficial effect on the increase of carotenoids content and amounted to 42% in β-carotene, 19% in lutein, and 14% in zeaxanthin. It confirms that modelling the content of carotenoids is possible using UVA and blue light (440 nm) during seed cultivation. An increase in the content of β-carotene and lutein in red clover sprouts was obtained in comparison to the test with white light without PEF pre-treatment, respectively by 8.5% and 6%. At the same time a 3.3% decrease in the content of zeaxanthin was observed. Therefore, it can be concluded that PEF pre-treatment may increase mainly the content of β-carotene.

Keywords: carotenoids; traditional food; light-emitting diodes; pulsed electric field; lutein; zeaxanthin; β-carotene

---

1. Introduction

Carotenoids play a key role in human health and should therefore be a regular part of the daily diet. It is important that they are delivered from various sources. Several carotenoids, including β- carotene, β-cryptoxanthin, and α-carotene are classified as provitamin A. Carotenoids other than provitamin A are also important in the human diet, because their high intake is correlated with a lower risk of developing chronic degenerative diseases including age-related, with special attention to age-related macular degeneration (AMD), cardiovascular diseases and some types of cancer [1–3].

Carotenoids are pigments soluble in lipids. Carotenoids are isoprenoid metabolites synthesized by all photosynthetic organisms (including algae, plants and even cyanobacteria) and some non-photosynthetic organisms such as archaea, fungi, bacteria or animals. Carotenoids are in over 1100 naturally occurring compounds that give colour to many edible parts of plants and flowers from yellow, through to orange and red. In addition, carotenoids can be cleaved to produce compounds with roles as growth regulators, such as abscisic acid (ABA) and strigolactones, as well as other bioactive molecules.

Carotenoids are found in many plant products. They are components of supplements and are also an additive to feed (e.g., lutein, zeaxanthin, β-cryptoxanthin, α-xanthine) in order to obtain the right colour of, for example, farmed fish eggs [3,4].

The literature on the subject has shown that, except for plants stained orange or red, what is associated with the colour of these compounds (e.g., papaya, carrots, peppers, also green plants such as sprouts) is a very good source of carotenoids [3]. In addition, sprouts belong to low-processed food. Their cultivation is fast (several days), easy and relatively cheap. However, there are no reports on the content of carotenoids in red clover sprouts. Commonly, carotenoids are associated with products that are from yellow to red, but these compounds are also commonly found in products containing chlorophylls. Carotenoids in chloroplasts help to absorb an excess of energy and dissipate it in the form of heat. In photosynthesis, carotenoids help to absorb light, but also play an important role in getting rid of solar energy excess. When a leaf is exposed to the full sun, it receives a large portion of energy. If this energy is not properly managed, it can destroy elements essential for photosynthesis. The seed germs are an interesting product containing carotenoids, obtained during just a few days of their cultivation. The germ is called an embryo that has pierced through the seed coat and developed a system (root and cotyledon) for self-feeding.

The European Union (EU) Commission Regulation 208/2013 from 11 March 2013 defines sprouts as a product obtained as a result of germination of seeds and their development in water or other carriers, which can be collected before the formation of proper leaves and intended for consumption as a whole, including seeds [5]. Sprouted seeds are one of the most nutritious and tastiest types of food in the world. Most sprouts can be eaten fresh. They are an addition or the base of salads, they are also suitable for many snacks and dishes, such as soups, dressings, dips and cocktails. Sprouts can enrich the nutritional value of many dishes, such as pizza, casseroles, croquettes, burgers, meat dishes, etc. Currently, the use of germinated seeds is unlimited and original desserts, smoothies and even sweets are prepared.

Sprouts in some especially Asian cuisines have been consumed for a long time and in Europe and the US they occupy an important place in vegetarian diets. Sprouted seeds, due to their composition, can be called a health bomb, because they contain nutrients and many compounds that have a positive effect on human health; they are primarily antioxidant compounds, including plant dyes such as carotenoids. The advantages of germinated seeds are primarily their nutritional and taste qualities, as well as their growing availability and diversity on the market. Seed cultivation can be carried out throughout the year using various solutions. Modern crops can be grown in phytotron (climatic) chambers with an automatic irrigation and temperature regulating system and additional lighting. In addition, assimilation lighting is one of the most important factors of plants grown in such conditions. Unlike photosynthesis, photoreaction is a more qualitative reaction and it depends on the wavelength. Thanks to this it is possible to stimulate photophysiological processes in the plant, affecting the content of bioactive compounds and obtaining their more often favourable composition.

To produce young edible seedlings (sprouts), it is possible to use the seeds of plants belonging, among others, to the *Fabaceae* family (bean beetles). The red clover (*Trifolium pratense L*) belongs to this family. Sprouted red clover seeds have strong antioxidant properties, mainly due to the high content of bioactive compounds. Studies have shown that they also contain a favourable qualitative and quantitative composition of phytoestrogenic compounds (isoflavones such as daidzein, genistein, formononetin and biochanin A), which have a beneficial effect on the human body. Sprouted red clover seeds can be included in the daily diet. In addition, light-emitting diodes (LEDs) open the possibility of their use for plant growth in a closed environment [6]. More advantageous and more efficient light sources are based on light-emitting diodes (LEDs) due to their advantages, such as light emission in a narrow spectrum band, high efficiency compared to traditional lamps, low voltage operation, photosynthetic regulation, photosynthetic photon flux density (PPFD) and low heat emission [7].

Plant growth, including germ and the profile of biologically active compounds, depends on the genotype, type of exposure (monochrome, combined or white light), its intensity and time [8]. The content of vitamins and microelements increases, and anti-nutritional components, such as trypsin inhibitors, are removed during germination as a result of intensive metabolism. It makes the germs safe for human consumption. The enzymes are activated in seed during germination, including amylolytic, proteolytic and lipolytic ones. Their activity favourably changes the composition of germinated seeds. Starch, proteins and fats are broken down, becoming a source of energy and substrates for the synthesis of new substances. Literature data [9] indicate the relationship between the content of antioxidant compounds in sprouts and their growth conditions (i.e., seed location, temperature and humidity). Seed cultivation can be carried out using many germination methods, not only differentiated by the method of moisturizing, temperature, and access to light, but also by the substrate [6,10,11]. Studies on the processes occurring in germinating seeds also explore the importance of light as an intermediary in the regulation of enzyme activity [12]. Seed viability can be expressed as the ability to germinate, which leads to the formation of a plant capable of reproduction. Sufficiently long viability affects the vigour of seeds, which expresses their ability to produce healthy and well-growing seedlings and plants. The cultivation of seeds both in terms of their germination efficiency and the content of bioactive compounds can be stimulated in various ways. An interesting factor is, for example, the usage of the pulsed electric field (PEF) on seeds before cultivation. The usage of the pulsed electric field on seeds is a phenomenon described in various ways in the literature. It has been shown that the effect of the PEF on seeds before germination has a positive effect on germination efficiency and seed growth rate [13,14]. Research was also conducted on the use of PEF [15–20], magnetic field [21] and UV light [22] in food processing and preservation. In biological material that has been subjected to an electric field, as a result of the impact of electrical impulses on cell membranes, a significant increase in their conductivity is observed, mainly through the formation of free spaces on the surface, the so-called pores. Their presence allows the free flow of various components through the cell membrane. Thanks to this phenomenon, it becomes possible to transfer ions, molecules and even more complex compounds (i.e., drugs, nucleic acids, monoclonal antibodies, oligonucleotides or plasmids) into the cell [23].

It is believed that the PEF can potentially be used to control and optimize the process of sprout growth and modify its composition, in particular nutritional values and bioactive ingredients [24]. PEF can also effectively stimulate germ growth and positively affect metabolism and nutrient content [24,25].

The effect of PEF on germination depends on the type of plant and the strength of the used field, while on some species such as marigold tomato or radish no significant effect was observed. In the case of lentil, a 50% increase in germination rate was observed. The effect of changes in germination rate induced by PEF is probably associated with changes in the metabolism of amino acids occurring in seeds [25] while PEF induces electroporation causing increased membrane permeability. Electroporation is a reversible process, however, when too high values of the PEF process parameters are used, irreversible changes in the structure of the cell membrane can occur [26]. Depending on the duration and the number of pulses and the strength of the electric field, the cell membrane may even be destroyed (i.e., irreversible electroporation) [25].

Red clover *Trifolium pratense* L. is an interesting plant. Throughout the world, it is most often known as a feed plant because it is used for fodder. There are many publications on bioactive compounds and their variability resulting from the cultivation of a mature plant. Many pharmaceutical preparations and dietary supplements [27,28] are also obtained on the basis of this plant. There is little research on carotenoids in red clover, and no such data were found in the available literature. Research and modelling of carotenoids content in red clover sprouts are innovative.

It should be noted that it has been confirmed that red clover sprouts are a rich source of bioactive compounds, including isoflavones, compounds with similar effects to oestrogen and can supplement the daily diet [6].

Another factor more and more often considered in the scientific literature, regarding the plants' cultivation, including sprouts, is the effect of light of different wavelengths on growth factors, as well as the content of biologically active compounds, including carotenoids [29]. Light, its intensity and wavelength have a significant impact on germination and plant development. Analysing the reports of various authors, the impact of light of varying wavelengths on sprouts of plants of different species is different. There is no information on the effect of light of varying wavelengths on the content of carotenoids in red clover sprouts. Therefore, the purpose of this work is to attempt to model the content of carotenoids in red clover sprouts using the effects of the PEF on their cultivation and the light of different wavelengths during their growth.

2. Materials and Methods

The experiment was carried out for seven days under different growing conditions. In the experiment, different wavelengths of light and pulsed electric fields (PEF) were used. The light-emitting diodes (LED) were used as the source of light. In the conducted research the light of wavelengths, ranging from 340–780 nm, was applied to the tested material. The strength of the pulsed electric field was chosen from the ranges 1 kV/cm, 2.5 kV/cm and 5 kV/cm. During the experiment, favourable conditions for seed development were created. The red clover seeds were put in a phytotron chamber where the light of different wavelengths was applied. Another group of seeds was treated with PEF before sowing and then they were cultivated in white light conditions. Cultivation of seeds was carried out to obtain the edible sprouts. The collection time was determined based on the quality features and content of bioactive compounds in red clover sprouts, including ascorbic acid and flavonoids (phytoestrogens). These compounds are present in the largest quantities between the 6th and 8th day of cultivation [6,11]. Thus, the germination in this experiment took place after 7 days of cultivation (after 168 h).

In the experiment, the sprouts were grown without soil and without supporting substrates containing mineral substances. Therefore, further development of the plant would be inhibited as a result of depletion of spare substances stored in seeds. Since whole sprouts of red clover are intended for consumption, the division into cotyledon and hypocotyl was not included in the study. It was found that on the day of harvesting red clover sprouts were firm, had green leaves with a cucumber-pea smell and were properly shaped, but depending on the cultivation conditions, the obtained weight (biomass) varied.

2.1. Pulsed Electric Field

A laboratory stand was built to conduct the experiment. It consisted of a high-voltage pulse generator with the voltage output in the range of 0 to 30 kV. The generated signal was the rectangular shape. The control system was used to obtain the proper value of the electric field strength. It allowed setting the number of pulses and the time interval between them. The process was conducted in the chamber in which the discharge occurred. There were two flat electrodes, between which a cylindrical, Teflon cell with seeds was placed.

The following parameters of the electroporation process were selected: pulse repeatability 10 s; the number of impulses 20; the electric field strengths 1 kV/cm, 2.5 kV/cm and 5 kV/cm. Higher field strength values were not used due to the destruction of the tested material. During the process, the constant value of temperature 20 °C was ensured. The temperature was measured with a thermocouple. The PEF was applied to red clover seeds before germination.

2.2. Raw Material

The seeds of red clover (*Trifolium pratense L.*), Rosette variety, suitable for germination supplied by FN Granum (Wodzierady, Poland) were chosen as the tested material. The plants were divided into 49 experimental series. Each series consisted of 5 g seeds. In each series there were three containers.

2.3. Sprouts Cultivation

Red clover (*Trifolium pratense* L.) seeds were grown using modified conditions. Seeds germination was carried out in the climatic chamber with a phytotron system (modified KBWF 720 Binder, Tuttlingen, Germany). Cultivation of the sprouts was carried out for 7 days at relative humidity 80% and 20 ± 1 °C. The LEDs which emitted the light with different wavelengths were used as the light sources in the process of sprouts growing. The different light wavelengths were used during cultivation: white light (380–780 nm), UVA (340 nm), blue (440 nm), red (630 nm). A photosynthetic photon flux density (PPFD) of 150 ± 5 μmol m^{-2} s^{-1} was maintained.

Since it is not possible to use sunlight in the phytotron chamber, white light (cool white) was used for all experimental samples to maintain constant growing conditions. Such lighting was also used by other authors in the cultivation of plants, including sprouts [30]. It would be possible to use sunlight in the experiment, but the sunlight would be variable in time and the environmental conditions would vary (e.g., in sprout machine).

In addition, the pulsed electric field (PEF) with three values of strength, 1, 2.5 and 5 kV/cm, respectively were applied on three series of sprouts before sowing. Seeds after PEF treatment were stored for 14 days at room temperature (T = 20 ° C) and then grown.

Sprouts treated with PEF were grown under white light (380–780 nm).

During cultivation, lighting was used repeatedly for 12 h and after that period the plants were stored 12 h in the dark. Sprouts were cultivated in portions of 5 g in containers (polypropylene) with a perforated bottom set on a tray. Sprouting was conducted in triplicate for each treatment (3 containers for each experiment). On the first day, the seeds were soaked by the addition of 15 mL of water, and in the subsequent days, the examined sprouts were irrigated to maintain high moisture in the culture. During cultivation, sprout weight growth was analysed every 24 h. Red clover sprouts were weighed immediately after harvesting on an analytical balance (RADWAG, PS 06.R2, Radom, Poland). The obtained germs of the sprouts are the average of three samples of 5 g seeds each. From the time the seeds were sown, the samples were weighed every 24 h.

The whole sprouts were examined. Sprouts were harvested manually every 24 h from the day of sowing. Analyses from the average sample from each container (3 containers for each treatment) were performed in one test samples taken from each container. After each harvest, the weight of the harvested sprouts was determined. The content of carotenoids was determined after 7 days of red clover sprouts growing.

These samples after harvesting were frozen immediately in liquid nitrogen and stored at −80 °C until further analysis.

2.4. Germination Energy

For each cultivation variant, four hundred seeds from the International Seed Testing Association (ISTA) (red clover and red clover after PEF treatment) were collected, grown in separate moulds made of polipropylen (PP) plastic, dedicated to conducting the germination process. During the study, the percentage of healthy, correctly germinated seeds was determined. During the cultivation, seed viability indexes were determined for energy of red clover seeds.

Germination energy (GE) was determined as the percentage of seeds that germinated during the first 4 days (96 h) [31].

2.5. Sample Preparation and Determination of Carotenoids

The samples of sprouts were placed in liquid nitrogen using an analytical mill (A 11 basic, IKA Works GmbH & Co. KG, Staufen, Germany)

The method proposed by Kimura and Rodriguez-Amaya [32] with some modifications was used for the extraction of carotenoids. First, 1 g of grounded and homogenized sprouts was placed in a plastic tube. Extraction was performed twice, first with 5 mL of acetone and next with 5 mL of ethanol (100%). Each extraction step was performed by vortex mixing for 2 min followed by centrifugation at 15,000 RCF (relative centrifugal force) for 10 min at 4 °C (MPW-360R, MPW Med. Instruments, Warsaw, Poland). The supernatants were evaporated to dryness under a stream of N_2 at 35 °C. The residue was redissolved in the mobile phase, centrifuged and diluted with the mobile phase to a suitable concentration.

The carotenoid extracts (injection volume 20 μL) were analysed on an HPLC system (Knauer Wissenschaftliche Geräte GmbH, Berlin, Germany) comprising of degasser of mobile phase Manager 5000, Pump 1000, autosampler model 3935 (maintained at 6 °C). A Gemini column maintained at 40 °C was used (5u C18110A, 150 × 4,60 mm 5 μm, Phenomenex, Torrance, CA, USA). The HPLC system was coupled with a photodiode array detector (PDA) model 2800. Quantification was performed at a detection wavelength of 445 nm. A gradient system (from A:B 75:25 to 0:100 in 13 min, maintaining this proportion until the end of the turn) was applied with methanol:water (90:10 *v/v*) as eluent A and acetonitrile:2-propanol (63:37 *v/v*) as eluent B. The flow rate was 1 mL/min. The carotenoids were identified based on the retention time of the standards and quantified with the help of peak areas against the standard calibration curves. Standards of carotenoids were purchased from Sigma-Aldrich (Saint Louis, MO, USA).

2.6. Statistical Analysis

Statistical analysis was based on the determination of the average values of three measurements and their standard deviation. The data were analysed using a one-way analysis of variance (ANOVA) with the Tukey's range test at the significance level $p < 0.05$. All data were tested for normality using the Shapiro–Wilk test. To test for homogeneity of variance, Levene's test was used. The calculations were performed using the software STATISTICA for Windows (version 10, Statsoft, Krakow, Poland).

3. Results and Discussion

In the experiment, a research hypothesis was put forward that it is possible to modify the content of carotenoids in red clover sprouts (*Trifolium pratense* L) by applying PEF to seeds before sowing and using light of varying wavelengths during cultivation. Until now, these factors have not been combined in such a way. Thus, this is an innovative approach to this issue. The highest sprout mass was obtained after applying red light to the seeds (38.9 g) and in the case of seeds cultivated in white light after PEF (5 kV/cm) pre-treatment (35.5 g). The lowest final mass was obtained in the case of UVA irradiation.

The seeds cultivation in white light after PEF pre-treatment with the field strengths of 1, 2.5 and 5 kV/cm increased the yield by 0.38%, 3.37% and 6.12%. Table 1 presents the final mass of the 100 g samples after 7 days of cultivation (on the day of harvest) in the case of different conditions: in UVA, blue, red and white light, without and after PEF pre-treatment with the field strengths of 1, 2.5 and 5 kV/cm. (Table 1). Statistically significant differences between the experimental series are presented in Table 2.

Table 1. Germination energy given per 100 pieces of seeds and the final mass of the samples after their cultivation in the different growth conditions, recalculated to the initial weight of 100 g sprouts. Values are presented as mean ± SD.

Light	Germination Energy (%) * (*n* = 4)	Final Mass (g) ** (*n* = 3)
UVA (340 nm)	89.3 ± 2.9	600.8 ± 36.6
blue (440 nm)	88.5 ± 2.1	629.0 ± 30.2
red (630 nm)	88.5 ± 2.3	778.6 ± 34.3
white (380–780 nm)	89.0 ± 2.3	669.5 ± 28.1
White Light and Pulsed Electric Field Strength	**Germination Energy (%) *** (*n* = 4)	**Final Mass (g) **** (*n* = 3)
1.0 kV/cm	88.8 ± 2.3	672.0 ± 21.7
2.5 kV/cm	90.8 ± 2.9	692.0 ± 20.8
5.0 kV/cm	94.0 ± 3.0	710.4 ± 30.5

* The performed analysis of variance does not allow for the rejection of the hypothesis of equality of means (F = 1.82, p = 0.143). Kruskal–Wallis test was conducted to examine the differences on germination energy according to the treatment taken. No significant differences (Chi-squared = 10.91, p = 0.09) were found among all the seven growing conditions. ** Statistically significant differences are presented in the Table 2.

Table 2. p-Values of Tukey's post hoc tests after one-way ANOVA for final mass of sprouts with treatments as fixed factors (F = 11.50, p < 0.05). Statistically significant differences are bolded.

Treatment	UVA	Blue	Red	White	PEF 1	PEF 2.5
UVA	-	-	-	-	-	-
blue	0.8931	-	-	-	-	-
red	**0.0002**	**0.0005**	-	-	-	-
white	0.1299	0.6335	**0.0066**	-	-	-
PEF 1	0.1094	0.5735	**0.0079**	1.0000	-	-
PEF 2.5	**0.0261**	0.1965	**0.0342**	0.9637	0.9796	-
PEF 5	**0.0063**	0.0522	0.1345	0.6254	0.6850	0.9827

Other authors also show a differentiated effect on the growth and yield of different plants depending on the wavelength of the used light. The results of the research indicate that the usage of red LED light in the cultivation of seeds and plants can cause an increase in biomass [33,34]. The red light is the most effective photosynthetically active radiation [35] referring to absorption. Research shows that blue light can also be used as a factor to increase biomass [36].

In the case of crops of different varieties of lettuce sprouts, it was found that the fresh weight increased in the case of using red or blue light in the growing process compared to crops under white light. The sprouts were irradiated for 12 h and left in the dark for 12 h. The best effect of the growth of germ mass was obtained when the sprouts were exposed to 3 h of blue light, and in one case it was observed that it increased by 23% [8]. On the other hand, in other studies [25] where wheat seeds before cultivation were treated with 50 pulses of 6 kV/cm PEF, a significant increase in fresh weight was noted, which is positively correlated with PEF energy.

A statistically significant effect of PEF (used at the initial stage of seed preparation for cultivation) on the mass of obtained red clover sprouts was observed. The exposure of seeds to the pulsed electric field can activate processes at an early stage of photomorphogenesis and thus positively affect its development. That line of research requires continuing experiments to confirm this observation. It was not the primary purpose of the presented studies.

In the case of wheatgrass (*Triticum aestivum* L.) the usage of 0.5 kV/cm PEF did not affect the growth of sprouts compared to the control ones. Increasing the PEF strength to 1.4 kV/cm had a positive effect on sprout growth, however, the 2 kV/cm PEF treatment of the tested seeds had an adverse effect and a lower growth was obtained than compared to the control [24].

Figure 1 shows the mass of sprouts depending on the day of cultivation. The initial mass of each sample was 5 g. The obtained mass in the harvest day depended on the pre-treatment and cultivation conditions.

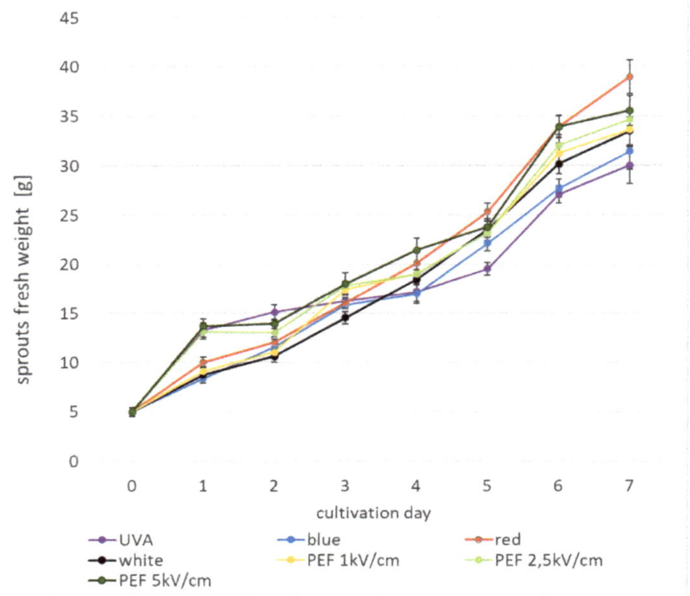

Figure 1. Mass of red clover sprouts during 7 days of sprout growth depending on irradiation and pulsed electric field (PEF) treatment prior to sowing. Values are presented as mean ± SD (*n* = 3).

The germination energy of red clover seeds (Table 1) in the case of PEF 5 kV/cm (in the process of seed pre-treatment before sowing) was the highest but did not differ statistically in comparison to other treatments. Moreover, all analysed groups had no significant differences in germination energy. The use of monochromatic LED sources and white light with the addition of 1 and 2.5 kV/cm PEF did not affect germination energy compared to white light crops (no statistically significant differences were found).

A statistically significant increase in germination energy was found after 4 days using a field of energy 240 J (12 kV) for parsley and 960 J (12 kV, the number of pulses was changed) for parsley, winter wheat, winter barley, lettuce, tomato and garden rocket [37].

In other studies, germination energy measured after 24 h was positively correlated with the increase in intensity (energy of pulse electric field) and in the case of 6 kV/cm PEF with 50 pulses it was 92% compared with 84% for non-treated sprouts [25].

For *Haloxsylon ammodendron* seeds, it has been observed that germination energy increases from 72.9% to 90.3% and 98.0% in the cases of using the strengths of 10 and 20 kV/cm, respectively [38].

It was observed that red clover sprouts after 7 days of cultivation contain β-carotene, lutein and zeaxanthin. However, their content varies depending on the factors used (PEF, different lengths of light). In this experiment, monochromatic light sources were used in the process of red clover sprouts growing. Moreover, the chosen wavelengths had a documented effect on photosynthesis and photomorphogenesis [36,39–41].

Green light was not chosen to process the experiment because it was documented in studies that a green monochromatic light source did not affect sprout development. Only slight enrichment of the spectrum with other light wavelengths was contemplated to improve photomorphogenesis [39,42].

In this study, it was observed that the content of carotenoids depends on the conditions of the used light (Figure 2). The dominant carotenoid in germinating red clover seeds is lutein, the content of which varies from 743 mg/kg in sprouts grown in red light, to 862 mg/kg in sprouts grown in UVA, and 888 mg/kg in sprouts grown in blue light. Lutein is a carotenoid with widely documented health-promoting properties. Obtaining new sources of lutein is desirable. At the same time, it was observed that the use of UVA or blue light during cultivation had a positive effect on the formation of other carotenoids, including β-carotene and zeaxanthin. UVA and blue light are the most preferred for obtaining carotenoid dyes. Under these cultivation conditions, the highest content of the total value of tested carotenoids (β-carotene, lutein and zeaxanthin) in red clover sprouts was obtained, respectively 1750 mg/kg (kg of dry weight) in the case of UVA irradiation and 1892 mg/kg using blue irradiation. Comparing these results to sprouts grown in white light, the increase in the content of these dyes in sprouts irradiated with blue light is statistically significant and equal to 42% in β-carotene, 19% in lutein and 14% in zeaxanthin (Figure 2, Tables 3–5). The content of lutein in sprouts from crops cultivated under red light was slightly lower than under white light, but the difference is not statistically significant (Table 4).

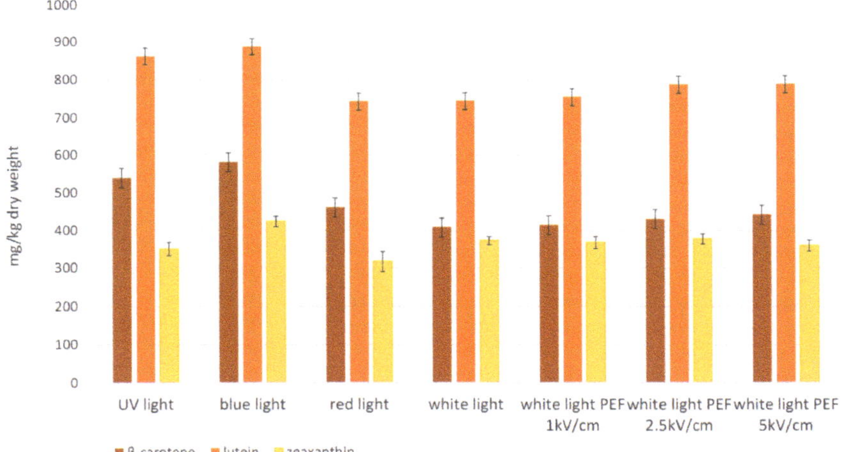

Figure 2. Influence of different light and PEF pre-treatment on carotenoids. The height of each bar and the error bars indicate the means and standard errors, respectively, from three independent measurements.

Table 3. *p*-Values of Tukey's post hoc tests after one-way ANOVA for β-carotene with treatments as fixed factors (F = 30.84, $p < 0.05$). Statistically significant differences are bolded.

Treatment	UVA	Blue	Red	White	PEF 1	PEF 2.5
UVA	-	-	-	-	-	-
blue	0.2244	-	-	-	-	-
red	**0.0056**	**0.0002**	-	-	-	-
white	**0.0002**	**0.0002**	0.0866	-	-	-
PEF 1	**0.0002**	**0.0002**	0.1591	0.9997	-	-
PEF 2.5	**0.0004**	**0.0002**	0.6018	0.8137	0.9470	-
PEF 5	**0.0010**	**0.0002**	0.9321	0.4290	0.6354	0.9911

Table 4. *p*-Values of Tukey's post hoc tests after one-way ANOVA for lutein with treatments as fixed factors (F = 21.10, $p < 0.05$). Statistically significant differences are bolded.

Treatment	UVA	Blue	Red	White	PEF 1	PEF 2.5
UVA	-	-	-	-	-	-
blue	0.7675	-	-	-	-	-
red	**0.0003**	**0.0002**	-	-	-	-
white	**0.0003**	**0.0002**	1.0000	-	-	-
PEF 1	**0.0006**	**0.0002**	0.9915	0.9966	-	-
PEF 2.5	**0.0133**	**0.0011**	0.2138	0.2511	0.5309	-
PEF 5	**0.0156**	**0.0013**	0.1859	0.2194	0.4801	1.0000

Table 5. *p*-Values of Tukey's post hoc tests after one-way ANOVA for zeaxanthin with treatments as fixed factors (F= 11.07, $p < 0.05$). Statistically significant differences are bolded.

Treatment	UVA	Blue	Red	White	PEF 1	PEF 2.5
UVA	-	-	-	-	-	-
blue	**0.0015**	-	-	-	-	-
red	0.2638	**0.0002**	-	-	-	-
white	0.6442	**0.0287**	**0.0151**	-	-	-
PEF 1	0.8147	**0.0163**	**0.0266**	0.9999	-	-
PEF 2.5	0.4240	0.0575	**0.0075**	0.9996	0.9903	-
PEF 5	0.9842	**0.0057**	0.0755	0.9669	0.9963	0.8470

Carotenoid content was studied in mature red clover plants for fodder grown in the field. Research shows that the main carotenoid is lutein, its content was 136 mg/kg DM (dry matter), and the total β-carotene content was 29 mg/kg DM [43]. In other studies, the composition of the carotenoids in red clover is different, and the content of lutein, β-carotene and zeaxanthin is much higher, respectively 237.7, 100.3 and 91.7 mg/kg DM. It should be noted that in recent studies, clover after harvesting has been subjected to a consolidation process [44].

Other authors report that the content of β-carotene in red clover is about 200 mg/kg DM, and the content of this carotenoid in *Trifolium repens* varies from 300 to 730 mg/kg DM depending on the harvesting place [45].

In the experiment where the subject was Alsike clover, also belonging to the genus *Trifolium*, the content of lutein and β-carotene was much higher and equalled 208.9–243 mg/kg and 35.4–123.4 mg/kg fresh weight, respectively [46].

The effect of various types of LED lighting on the content of carotenoids in alfalfa sprouts (also belonging to *Fabaceae* family) was investigated [47]. The combination of the red and blue LEDs used during cultivation increased lutein content from 82.6 to 108.2, and β-carotene from 26.6 to 44.6 mg/kg fresh mass of sprouts [47]. However, the research did not cover the whole sprout, only cotyledon. Increasing the intensity of blue light during the cultivation of beetroot (*Beta vulgaris* L.) and parsley (*Petroselinum crispum* Mill.) caused an increase in the content of carotenoids as compared to irradiation with red light. It has been shown that increasing the proportion of blue light intensity in these microgreens results in an increase in lutein content from 103.8 to 118.5 mg/kg of fresh beet mass, from 122.9 to 190.7 mg/kg of fresh parsley, and β-carotene, respectively from 0.09 to 0.87 and 0.54 to 0.86 mg/kg, and zeaxanthin from 1.39 to 3.20 and up to 0.84–14.4 mg/kg, respectively [35].

Lefsrud in his research observed an increase in the content of lutein and β-carotene in kale when illuminated with both blue and red light. The maximum accumulation of lutein and β-carotene in fresh kale mass was found at 640 nm and 440 nm, respectively, calculated on a fresh mass basis. However, when converted to dry mass, the maximum lutein content in kale also occurred at 440 nm. [1].

There are no reports in the literature on the content of carotenoids in red clover sprouts, as well as on the impact of PEFs on the content of carotenoids. Therefore, our research can only be compared with the unique research carried out by Ahmed et al. [25] who studied the effect of PEF on the content of plant dyes in wheat germ. A 35% increase in the content of carotenoids was observed in the PEF treatment with strength of 6 kV/cm and 50 impulses compared to untreated sprouts [25].

In our experiment, no such significant effect of PEF treatment of red clover seeds on the content of carotenoids was found.

Modelling the content of carotenoids in sprouts which are grown in different lighting conditions, especially in ultraviolet radiation with wavelengths up to 400 nm and blue light, forces the plant to adapt to the changes to avoid their harmful effects. Many compounds (e.g., flavonoids as well as carotenoids) can fulfil the role of protection against the harmful effects of UV light [6,48]. Carotenoids are the auxiliary photoreceptors of chlorophyll and absorb light mainly in the blue region. Cryptochromes are the receptors of blue light in the 390–480 nm range. They stimulate leaf expansion. In the blue and UVA spectrum, phototropins, responsible for plant phototropism, bending of shoots towards the light, opening of stomata and leaf expansion are also photoreceptors in the blue and UVA spectrum [49,50].

The increase of carotenoids content is plants' response to stress is associated with high irradiance. High exposure does not always lead to the growth of carotenoids, it can sometimes cause photodegradation of pigment particles [1,51,52]

In the presented experiments, we observed that monochromatic blue light increases the content of carotenoids. The positive effect of blue light on the content of these dyes was also observed by other authors [52,53] Opposite observations were made by Tuan et al. [29]. Research on tartary buckwheat sprouts has shown that blue light causes a decrease in the content of carotenoids compared to white light. However, it should be noticed that in the recalled work, the full spectrum range is not given in the description of the characteristics of the white light source (only 380 nm is mentioned (i.e., UV wavelength)).

The effect of pulsed electric field on seeds (before cultivation) had an impact on the increase of carotenoid content, including β-carotene, lutein and zeaxanthin. It was observed that the application of 5 kV/cm had the most favourable effect on the increase in the content of carotenoids in red clover sprouts compared with the reference test which grew seeds in white light. An increase in the content of β-carotene and lutein in red clover sprouts was obtained in comparison to the test with white light and without PEF pre-treatment by 8.5% and 6%, respectively, and a decrease in the content of zeaxanthin by 3.3% was also found. Therefore, PEF pre-treatment may increase mainly the content of β-carotene. At the same time, it should not be assumed that the use of a higher value of electric field strength (above 5 kV/cm) will work more favourably. The use of higher field strength can damage or destroy the seed epidermis and damage its internal structure, which means that the seed will not be able to germinate. The results of this experiment indicate that the PEF pre-treatment on seeds before their cultivation can be one of the factors that can model the carotenoids content in germinated seeds. The obtained results should be considered as an introduction to further research.

Photosynthetically active radiation includes a wide range of light wavelengths that do not participate in the process of photosynthesis but can stimulate the vegetative growth of plants and modify the chemical composition of the leaves [54].

4. Conclusions

The dominant carotenoid in germinating red clover seeds is lutein, whose content varies from 743 mg/kg in sprouts grown in red light, 862 mg/kg in sprouts grown in UVA, to 888 mg/kg in sprouts grown in blue light. UVA and blue light are the most preferred for obtaining carotenoid dyes. The highest content of the total value of tested carotenoids (β-carotene, lutein and zeaxanthin) in red clover sprouts was obtained, respectively at 1750 mg/kg (kg of dry weight) in the case of UVA irradiation and 1892 mg/kg using blue irradiation.

Modelling the content of carotenoids in red clover sprouts is possible because by using UVA and blue light (440 nm) during seed cultivation, a significant increase in β-carotene, lutein and zeaxanthin is obtained. An interesting result of the presented research is also the use of PEF pre-treatment on the seeds before the cultivation process. It is a factor that, although to a lesser extent than blue light and

UVA, also causes an increase of the carotenoids in sprouts. It is possible that PEF pre-treatment of seeds will also have a beneficial effect on other seeds intended for germination. It requires further testing and confirmation. At the same time, the use of red light resulted in a greater mass of red clover sprouts, because red light has a beneficial effect on photomorphogenesis.

Consumption of 100 g fresh red clover sprouts which were grown for 7 days in UVA, blue, red or white light provided 14.0, 15.1, 12.1 and 12.2 mg total sum of tested carotenoids, respectively. Although the PEF treatment does not significantly affect the carotenoids' content, in the case of PEF strength 5 kV/cm the intake of carotenoids increased by 4% compared to cultivation only in white light and equals 12.7 mg/100 g.

It is worth noting that red clover sprouts may also be a new source of carotenoids including lutein and zeaxanthin. The recommended dietary allowances (RDA) of lutein and zeaxanthin in the human diet have not been established. For reducing the risk of AMD, the efficacious intake level for lutein may be ~ 6 mg per day [55]. Therefore, the consumption of approx. 40 g of red clover sprouts may cover the daily requirement.

Cultivation of seeds for sprouts in closed conditions (in a phytotron chamber) with properly selected blue or UVA lighting can stimulate an increase in the content of bioactive compounds, including carotenoids, and at the same time ensure the proper development of leaves. Changing the spectral composition of radiation during cultivation can be used to produce plants intended for several days of germination and consumption. It is possible to obtain a significantly improved product in terms of nutritional value in a short time using the tested conditions.

Author Contributions: Conceptualization, I.G.-C. and E.K.; methodology, I.G.-C., E.K. and P.K.; validation, I.G.-C., E.K., A.C., P.K. and T.D.; formal analysis, I.G.-C. and E.K.; investigation, I.G.-C., E.K., A.C., P.K. and T.D.; resources, I.G.-C. and P.K. and T.D.; data curation, I.G.-C., P.K. and T.D.; writing—original draft preparation, I.G.-C., E.K. and A.C.; writing—review and editing, I.G.-C., E.K. and A.C.; visualization, A.C.; supervision, I.G.-C.; project administration, I.G.-C. and E.K.; funding acquisition, I.G.-C.; P.K. and T.D. All authors have read and agreed to the published version of the manuscript.

Funding: This research received no external funding.

Conflicts of Interest: The authors declare no conflicts of interest.

References

1. Lefsrud, M.G.; Kopsell, D.; Kopsell, D.E.; Curran-Celentano, J. Irradiance levels affect growth parameters and carotenoid pigments in kale and spinach grown in a controlled environment. *Physiol. Plant.* **2006**, *127*, 624–631. [CrossRef]

2. Eggersdorfer, M.; Wyss, A. Carotenoids in human nutrition and health. *Arch. Biochem. Biophys.* **2018**, *652*, 18–26. [CrossRef]

3. Concepción, M.R.; Avalos, J.; Bonet, M.L.; Boronat, A.; Gómez-Gómez, L.; Hornero-Méndez, D.; Limón, M.C.; Meléndez-Martínez, A.J.; Olmedilla-Alonso, B.; Palou, A.; et al. A global perspective on carotenoids: Metabolism, biotechnology, and benefits for nutrition and health. *Prog. Lipid Res.* **2018**, *70*, 62–93. [CrossRef]

4. Gałązka-Czarnecka, I.; Korzeniewska, E.; Czarnecki, A.; Sójka, M.; Kiełbasa, P.; Dróżdż, T. Evaluation of Quality of Eggs from Hens Kept in Caged and Free-Range Systems Using Traditional Methods and Ultra-Weak Luminescence. *Appl. Sci.* **2019**, *9*, 2430. [CrossRef]

5. Official Journal of the European Union. Commission Implementing Regulation (EU) No 208/2013 of 11 March 2013 on traceability requirements for sprouts and seeds intended for the production of sprouts. *Off. J. Eur. Union* **2013**, *68*, 16.

6. Budryn, G.; Gałązka-Czarnecka, I.; Brzozowska, E.; Grzelczyk, J.; Mostowski, R.; Żyżelewicz, D.; Cerón-Carrasco, J.P.; Pérez-Sánchez, H. Evaluation of estrogenic activity of red clover (Trifolium pratense L.) sprouts cultivated under different conditions by content of isoflavones, calorimetric study and molecular modelling. *Food Chem.* **2017**, *245*, 324–336. [CrossRef]

7. Morrow, R.C. LED Lighting in Horticulture. *HortScience* **2008**, *43*, 1947–1950. [CrossRef]

8. Pardo, G.P.; Aguilar, C.H.; Martínez, F.R.; Pacheco, A.D.; Martínez, C.L.; Ortiz, E.M. High intensity led light in lettuce seed physiology (Lactuca sativa L.). *Acta Agroph.* **2013**, *204*, 665–677.

9. Guzmán-Ortiz, F.A.; Castro-Rosas, J.; Gómez-Aldapa, C.A.; Mora-Escobedo, R.; Rojas-León, A.; Rodriguez-Marin, M.L.; Falfán-Cortés, R.N.; Román-Gutiérrez, A.D. Enzyme activity during germination of different cereals: A review. *Food Rev. Int.* **2018**, *35*, 177–200. [CrossRef]

10. Korzeniewska, E.; Gałązka-Czarnecka, I.; Czarnecki, A. Influence of thin silver layers applied in physical vacuum deposition on energy and sprouting ability of plant seeds from the Fabaceae family. *Prz. Elektrotech.* **2020**, *96*, 250–253.

11. Brzozowska, E.; Gałązka-Czarnecka, I.; Krala, L. Effect of diffuse solar radiation on selected properties of red cover sprouts (*Trifolium pratense* L). *Żywn. Nauka Technol. Jakość* **2014**, *6*, 67–80. [CrossRef]

12. Kim, E.H.; Kim, S.H.; Chung, J.I.; Chi, H.Y.; Kim, J.A.; Chung, I.M. Analysis of phenolic compounds and isoflavones in soybean seeds (Glycine max (L.) Merill) and sprouts grown under different conditions. *Eur. Food Res. Technol.* **2005**, *222*, 201–208. [CrossRef]

13. Eing, C.; Bonnet, S.; Pacher, M.; Puchta, H.; Frey, W. Effects of nanosecond pulsed electric field exposure on arabidopsis thaliana. *IEEE Trans. Dielectr. Electr. Insul.* **2009**, *16*, 1322–1328. [CrossRef]

14. Starodubtseva, G.P.; Livinski, S.A.; Gabriyelyan, S.Z.; Lubaya, S.I.; Afanace, M.A. Process Control of Pre-Sowing Seed Treatment by Pulsed Electric Field. In *Acta Technologica Agriculturae 1*; Slovaca Universitas Agriculturae Nitriae: Nitra, Slovakia, 2018; pp. 28–32. [CrossRef]

15. Gocławski, J.; Sekulska-Nalewajko, J.; Korzeniewska, E.; Piekarska, A. The use of optical coherence tomography for the evaluation of textural changes of grapes exposed to pulsed electric field. *Comput. Electron. Agric.* **2017**, *142*, 29–40. [CrossRef]

16. Dziadek, K.; Kopeć, A.; Dróżdż, T.; Kiełbasa, P.; Ostafin, M.; Bulski, K.; Oziembłowski, M. Effect of pulsed electric field treatment on shelf life and nutritional value of apple juice. *J. Food Sci. Technol.* **2019**, *56*, 1184–1191. [CrossRef]

17. Ostafin, M.; Bulski, K.; Dróżdż, T.; Nawara, P.; Nęcka, K.; Lis, S.; Kiełbasa, P.; Tomasik, M.; Oziembłowski, M. The influence of the AC electromagnetic field on growth of Yarrowia lipolytica yeast. *Prz. Elektrotech.* **2016**, *12*, 117–121.

18. Korzeniewska, E.; Gałązka-Czarnecka, I.; Czarnecki, A.; Piekarska, A.; Krawczyk, A. Influence of PEF on antocyjans in wine. *Prz. Elektrotech.* **2018**, *94*, 57–60.

19. Gałązka-Czarnecka, I.; Korzeniewska, E.; Czarnecki, A.; Politowski, K. Modification of polyphenol content in wines using pulsed electric field. *Prz. Elektrotech.* **2019**, *95*, 89–92.

20. Iatcheva, I.; Saykova, I. FEM Modelling of Heat Effects in Cellular Materials under the Application of Electric Fields. In Proceedings of the 19th International Symposium on Electromagnetic Fields in Mechatronics, Electrical and Electronic Engineering, (ISEF), Nancy, France, 29–31 August 2019; IEEE: Piscataway, NJ, USA; pp. 1–2. [CrossRef]

21. Jakubowski, T. The effect of stimulation of seed potatoes (Solanum tuberosum L.) in the magnetic field on selected vegetation parameters of potato plants. *Prz. Elektrotech.* **2020**, *96*, 166–169.

22. Sobol, Z.; Jakubowski, T. The effect of storage duration and UV-C stimulation of potato tubers, and soaking of potato strips in water on the density of intermediates of French fries production. *Prz. Elektrotech.* **2020**, *1*, 244–247. [CrossRef]

23. Kotulska, M. Electrochemotherapy in cancer treatment. *Adv. Clin. Exp. Med.* **2007**, *16*, 601–607.

24. Leong, S.Y.; Burritt, D.J.; Oey, I. Electropriming of wheatgrass seeds using pulsed electric fields enhances antioxidant metabolism and the bioprotective capacity of wheatgrass shoots. *Sci. Rep.* **2016**, *6*, 25306. [CrossRef] [PubMed]

25. Ahmed, Z.; Manzoor, M.F.; Ahmad, N.; Zeng, X.; Din, Z.U.; Roobab, U.; Qayum, A.; Siddique, R.; Siddeeg, A.; Rahaman, A. Impact of pulsed electric field treatments on the growth parameters of wheat seeds and nutritional properties of their wheat plantlets juice. *Food Sci. Nutr.* **2020**, *8*, 2490–2500. [CrossRef] [PubMed]

26. Saulis, G. Electroporation of Cell Membranes: The Fundamental Effects of Pulsed Electric Fields in Food Processing. *Food Eng. Rev.* **2010**, *2*, 52–73. [CrossRef]

27. Bemis, D.L.; Capodice, J.L.; Costello, J.E.; Vorys, G.C.; Katz, A.E.; Buttyan, R. The use of herbal and over-the-counter dietary supplements for the prevention of prostate cancer. *Curr. Oncol. Rep.* **2006**, *8*, 228–236. [CrossRef] [PubMed]

28. Tundis, R.; Marrelli, M.; Conforti, F.; Tenuta, M.C.; Bonesi, M.; Menichini, F.; Loizzo, M.R. Trifolium pratense and T. repens (Leguminosae): Edible Flower Extracts as Functional Ingredients. *Foods* **2015**, *4*, 338–348. [CrossRef]

29. Tuan, P.A.; Thwe, A.A.; Kim, Y.B.; Kim, J.K.; Kim, J.-B.; Lee, S.; Chung, S.-O.; Park, S.U. Effects of White, Blue, and Red Light-Emitting Diodes on Carotenoid Biosynthetic Gene Expression Levels and Carotenoid Accumulation in Sprouts of Tartary Buckwheat (Fagopyrum tataricum Gaertn.). *J. Agric. Food Chem.* **2013**, *61*, 12356–12361. [CrossRef]

30. Wu, M.-C.; Hou, C.-Y.; Jiang, C.-M.; Wang, Y.-T.; Wang, C.-Y.; Chen, H.-H.; Chang, H.-M. A novel approach of LED light radiation improves the antioxidant activity of pea seedlings. *Food Chem.* **2007**, *101*, 1753–1758. [CrossRef]

31. International Seed Testing Association. *International Rules for Seed Testing*; International Seed Testing Association (ISTA): Bassersdorf, Switzerland, 2009.

32. Kimura, M.; Amaya, D.B.R. A scheme for obtaining standards and HPLC quantification of leafy vegetable carotenoids. *Food Chem.* **2002**, *78*, 389–398. [CrossRef]

33. Heo, J.-W.; Kang, D.-H.; Bang, H.-S.; Hong, S.; Chun, C.-H.; Kang, K.-K. Early Growth, Pigmentation, Protein Content, and Phenylalanine Ammonia-lyase Activity of Red Curled Lettuces Grown under Different Lighting Conditions. *Korean J. Hortic. Sci. Technol.* **2012**, *30*, 6–12. [CrossRef]

34. Zhang, T.; Shi, Y.; Piao, F.; Sun, Z. Effects of different LED sources on the growth and nitrogen metabolism of lettuce. *Plant Cell Tissue Organ Cult.* **2018**, *134*, 231–240. [CrossRef]

35. Terashima, I.; Fujita, T.; Inoue, T.; Chow, W.S.; Oguchi, R. Green Light Drives Leaf Photosynthesis More Efficiently than Red Light in Strong White Light: Revisiting the Enigmatic Question of Why Leaves are Green. *Plant Cell Physiol.* **2009**, *50*, 684–697. [CrossRef] [PubMed]

36. Li, Y.; Zheng, Y.; Liu, H.; Zhang, Y.; Hao, Y.; Song, S.; Lei, B. Effect of supplemental blue light intensity on the growth and quality of Chinese kale. *Hortic. Environ. Biotechnol.* **2018**, *60*, 49–57. [CrossRef]

37. Evrendilek, G.A.; Karatas, B.; Uzuner, S.; Tanaso, I. Design and effectiveness of pulsed electricfields towards seed disinfection. *J. Sci. Food Agric.* **2019**, *99*, 3475–3480. [CrossRef] [PubMed]

38. Su, B.; Guo, J.; Nian, W.; Feng, H.; Wang, K.; Zhang, J.; Fang, J. Early Growth Effects of Nanosecond Pulsed Electric Field (nsPEFs) Exposure on Haloxylon ammodendron. *Plasma Process. Polym.* **2014**, *12*, 372–379. [CrossRef]

39. Bugbee, B. Toward an optimal spectral quality for plant growth and development: The importance of radiation capture. *Acta Hortic.* **2016**, *1134*, 1–12. [CrossRef]

40. Frede, K.; Schreiner, M.; Baldermann, S. Light quality-induced changes of carotenoid composition in pak choi Brassica rapa ssp. chinensis. *J. Photochem. Photobiol. B Biol.* **2019**, *193*, 18–30. [CrossRef]

41. Samuolienė, G.; Viršilė, A.; Brazaitytė, A.; Jankauskienė, J.; Sakalauskienė, S.; Vaštakaitė-Kairienė, V.; Novičkovas, A.; Viškelienė, A.; Sasnauskas, A.; Duchovskis, P. Blue light dosage affects carotenoids and tocopherols in microgreens. *Food Chem.* **2017**, *228*, 50–56. [CrossRef]

42. Folta, K.M.; Maruhnich, S.A. Green light: A signal to slow down or stop. *J. Exp. Bot.* **2007**, *58*, 3099–3111. [CrossRef]

43. Cardinault, N.; Doreau, M.; Poncet, C.; Nozière, P. Digestion and absorption of carotenoids in sheep given fresh red clover. *Anim. Sci.* **2006**, *82*, 49–55. [CrossRef]

44. Chauveau-Duriot, B.; Thomas, D.; Portelli, J.; Doreau, M. Effect du mode de conservation sur la teneur en caroténoïdes des fourrages Carotenoid content in forages: Variation during conservation. In *12. Rencontres Autour des Recherches sur les Ruminants*; Institut de l'Elevage: Paris, France, 2005; Volume 12, p. 117.

45. Mikhailov, A.L.; Timofeeva, O.A.; Ogorodnova, U.A.; Stepanov, N.S. Comparative Analysis of Biologically Active Substances in Trifolium Pratense and Trifolium Repens Depending on the Growing Conditions. *J. Environ. Treat. Tech.* **2019**, *7*, 874–877.

46. Zeb, A.; Hussain, A. Chemo-metric analysis of carotenoids, chlorophylls, and antioxidant activity of Trifolium hybridum. *Heliyon* **2020**, *6*, e03195. [CrossRef]

47. Fiutak, G.; Michalczyk, M.; Filipczak-Fiutak, M.; Fiedor, L.; Surówka, K. The impact of LED lighting on the yield, morphological structure and some bioactive components in alfalfa (Medicago sativa L.) sprouts. *Food Chem.* **2019**, *285*, 53–58. [CrossRef] [PubMed]

48. Schmidt, J.; Batic, F.; Pfanz, H. Photosynthetic performance of leaves and twigs of evergreen holly (Ilex aquifolium L.). *Phyton* **2000**, *40*, 179–190.

49. Ahmad, M.; Grancher, N.; Heil, M.; Black, R.C.; Giovani, B.; Galland, P.; Lardemer, D. Action Spectrum for Cryptochrome-Dependent Hypocotyl Growth Inhibition in Arabidopsis1. *Plant Physiol.* **2002**, *129*, 774–785. [CrossRef]

50. Sancar, A. Structure and function of DNA photolyase and cryptochrome blue-light photoreceptors. *Chem. Rev.* **2003**, *103*, 2203–2238. [CrossRef]
51. Lee, S.-H.; Tewari, R.K.; Hahn, E.-J.; Paek, K.-Y. Photon flux density and light quality induce changes in growth, stomatal development, photosynthesis and transpiration of Withania Somnifera (L.) Dunal. plantlets. *Plant Cell Tissue Organ Cult.* **2007**, *90*, 141–151. [CrossRef]
52. Kopsell, D.; Pantanizopoulos, N.I.; Sams, C.; Kopsell, D.E. Shoot tissue pigment levels increase in 'Florida Broadleaf' mustard (Brassica juncea L.) microgreens following high light treatment. *Sci. Hortic.* **2012**, *140*, 96–99. [CrossRef]
53. Zhang, L.; Ma, G.; Yamawaki, K.; Ikoma, Y.; Matsumoto, H.; Yoshioka, T.; Ohta, S.; Kato, M. Effect of blue LED light intensity on carotenoid accumulation in citrus juice sacs. *J. Plant Physiol.* **2015**, *188*, 58–63. [CrossRef]
54. Olle, M.; Viršilė, A. The effects of light-emitting diode lighting on greenhouse plant growth and quality. *Agric. Food Sci.* **2013**, *22*, 223–234. [CrossRef]
55. Seddon, J.M.; Ajani, U.A.; Sperduto, R.D.; Hiller, R.; Blair, N.; Burton, T.C.; Farber, M.D.; Gragoudas, E.S.; Haller, J.; Miller, D.T.; et al. Dietary Carotenoids, Vitamins A, C, and E, and Advanced Age-Related Macular Degeneration. *JAMA* **1994**, *272*, 1413–1420. [CrossRef] [PubMed]

© 2020 by the authors. Licensee MDPI, Basel, Switzerland. This article is an open access article distributed under the terms and conditions of the Creative Commons Attribution (CC BY) license (http://creativecommons.org/licenses/by/4.0/).

Article

Bunch Microclimate Affects Carotenoids Evolution in cv. Nebbiolo (*V. vinifera* L.)

Andriani Asproudi [1], Maurizio Petrozziello [1,*], Silvia Cavalletto [2], Alessandra Ferrandino [2], Elena Mania [2] and Silvia Guidoni [2]

[1] Consiglio per la ricerca in agricoltura e l'analisi dell'economia agraria (CREA). Via P. Micca 35, 14100 Asti, Italy; andriani.asproudi@crea.gov.it

[2] Dipartimento di Scienze Agrarie, Forestali e Alimentari (Università di Torino); Largo Braccini 2, 10095 Grugliasco, Italy; silvia.cavalletto@unito.it (S.C.); alessandra.ferrandino@unito.it (A.F.); elena.mania@unito.it (E.M.); silvia.guidoni@unito.it (S.G.)

* Correspondence: maurizio.petrozziello@crea.gov.it; Tel.: +39-0141433820

Received: 30 April 2020; Accepted: 29 May 2020; Published: 31 May 2020

Abstract: This study investigates the impact of bunch microclimate on the evolution of some relevant carotenoids in Nebbiolo grapes. Four bunch-zone microclimates, defined by different vineyard aspect and vine vigor, were characterized by radiation and temperature indices. Berry samples were collected from green phase up to harvest, during two consecutive seasons and carotenoid determination was assessed by High-Performance Liquid Chromatography (HPLC). High carotenoid concentrations were highlighted in Nebbiolo. Lutein and neoxanthin contents (μg berry^{-1}) varied similarly in both seasons achieving a concentration peak after veraison especially in the cooler plots while a variety effect on the lutein seasonal trend was presumed. Conversely, β-carotene content remained generally constant during ripening, with the exception of the south plots showing dissimilar evolution between the seasons. Furthermore, higher temperature in the less vigorous and south facing vineyards led to lower amounts of carotenoids, both during ripening and at harvest. Bunch zone temperature and light condition may affect both synthesis and degradation of grape carotenoids determining their amount and profile at harvest. These findings add further knowledge about the influence of climate changes on grape aroma precursors, and are useful to adapt cultural strategies and preserve grape quality consequently.

Keywords: vineyard aspect; vineyard topography; vine vigor; heat accumulation; temperature; photosynthetically active radiation; lutein; neoxanthin; β-carotene

---

1. Introduction

Plant carotenoids are essential for photosynthesis and photoprotection due to their multiple functions as potent free radical quenchers, singlet oxygen scavengers and lipid antioxidants. They are present in the photosynthetic tissues as part of photosystem II [1]. Carotenoids also give rise to the formation of numerous biologically active cleavage products such as aroma compounds, vitamins, phytohormones, and apocarotenoid pigments [2].

Grape carotenoids were identified as precursors of certain key odorants in wine, namely C_{13}-norisoprenoids, which are low threshold aroma compounds characterized by floral and fruity pleasant notes strongly linked to increases in wine quality, especially for non-floral varieties [3]. The formation of norisoprenoids is thought to occur from the biodegradation of the parent carotenoid, followed by enzymatic conversion to the aroma precursor (e.g., a glycosylated or other polar intermediate), and finally by the acid-catalyzed conversion to the aroma compound [4], which may be then subjected to further acid reaction during wine aging [5]. A family of region-specific carotenoid cleavage dioxygenase (CCD) enzymes is implicated in the initial biodegradation and oxidative cleavage of carotenoids to form plant apocarotenoids, e.g., C13-norisoprenoids [6–8]. The expression of a CCD capable of producing C13-norisoprenoids from lutein and zeaxanthin (*VvCCD1*) was reported to increase at veraison [6], while the reported increase in expression of a CCD4 gene (*VvCCD4*) after veraison is suggestive of a possible role of this enzyme on norisoprenoid formation during the late stage of berry ripening [8]. Carotenoids could also be precursors of norisoprenoids during fermentation and wine aging [9–11].

Lutein and β-carotene represent nearly 85% of the total carotenoids in grapes and they are mostly involved in degradation reactions in grapes, juice, and wine. The carotenoids directly involved in the aroma of wine are β-carotene, generating β-ionone, and neoxanthin generating β-damascenone. Lutein and violaxanthin also undergo breakdown reactions that may produce norisoprenoid compounds in wines [3,12]. Lutein, for instance, is reported to be an important precursor of 1,1,6-trimethyl-1,2-dihydronaphthalene (TDN) [13] while the formation of megastigmane-3,9-diol and of 3-oxo-α-ionol glucosides from the ε-cycle of lutein has also been illustrated [14]. Moreover, the involvement of lutein epoxide (Lx) cycle, an additional slower and reversible mechanism of photoprotection that supplements the violaxanthin cycle (violaxanthin and zeaxanthin), was recently demonstrated [8,15]. Authors observed a higher accumulation of lutein 5,6-epoxide in shaded berries [15] that is de-epoxidized to lutein following normal light aspect [8]. Because of the activation of the xanthophyll and Lx cycles at the end of the ripening period [14], processes of bioconversion between different carotenoids, also induced by modified light conditions, may take place, and influence the formation of norisoprenoids [6,14,16].

Generally, carotenoids are thought to be mostly synthesized between berry set and veraison. For this reason, the aromatic profile of the wine also depends on the carotenoid composition of immature berries while the end of veraison (and not the beginning as thought in the past) appears to be a key moment for the changes in the ratio carotenoid norisoprenoid [17]. Several variables may promote the norisoprenoid final content in grapes: some favor carotenoid synthesis during the herbaceous phase of berry growth, some others stimulates their degradation to norisoprenoids occurring from veraison onwards [16,18]. Therefore, carotenoid evolution during ripening can be considered as an indicator of grape aromatic potential [2,19,20].

As reported in many studies, the concentration of carotenoids in ripe grapes depends on cultivars [21], ripening stage, climate region [22], altitude [23], soil water retention capacity [21], and degree of bunch exposure to sunlight [15]. The highest carotenoid concentration occurred in the hot regions, likely due to the higher amount and intensity of the received solar radiation [22,24]. In warm climates, the level of β-carotene at harvest resulted as three–six-fold higher than that of lutein [18,19,25]. In particular, light is the main factor responsible for the changes in the biosynthesis of carotenoids [4] promoting both their accumulation before veraison and causing their degradation during ripening [26]. The degree of the bunch exposure to sunlight appeared to influence the ratio epoxyxanthophylls: non-epoxyxanthophylls whereas high UV-B levels favored carotenoid degradation [23,27], actually, higher rates of degradation emerged during the hotter period of grape ripening [28].

Furthermore, vineyard topographic features, such as slope gradient and aspect or altitude, along with the heterogeneity of vineyard vigor and different vine management may generate a great variability in microclimatic conditions (light, temperature, and humidity) within vineyards, canopy and bunch zone, likely influencing grape ripening and quality. The impact of the vineyard microclimatic characteristics on Nebbiolo grape development, ripening and anthocyanin accumulation, as well as on the evolution of grape norisoprenoid precursors has already been investigated [29–31]. Until now, only one study regarding the evolution of the carotenoid compounds in Nebbiolo grapes has been carried out [28].

Many studies explored the impact of artificial regulation of the bunch exposure to sunlight, by leaf removal or other canopy manipulation, on grape metabolic composition [15,22,23,32]. Nevertheless, integrated studies focusing on the impact of vineyard aspect and natural vine vigor on bunch microclimate and grape aroma precursors are lacking.

Therefore, the aim of this study was to complete a previous research by assessing the concentration and seasonal accumulation pattern of the most relevant carotenoids in Nebbiolo grapes as well as to study the link between bunch microclimate and carotenoid evolution during ripening.

2. Materials and Methods

2.1. Vineyard Site and Treatments

This study is complementary of a previous research on Nebbiolo grapes [31]. The experimental vineyard, site and treatments are widely described in the cited article. Briefly, the study was performed during 2012 and 2013 in two commercial vineyards located in North-West Italy (44°36′04″ N, 8°00′34″ E; 428 m above sea level). Four vineyards differing in terms of slope aspect (South and West) and natural vine vigor (two level of vigor in each vineyard: V+ and V−) were identified by assessing the Normalized Difference Vegetation Index of the parcel, as previously described [33]. In 2012, the vineyards compared were SouthV+, SouthV−, WestV+; in 2013, the WestV− vineyard was included. In each vineyard, three replicates of 50 vines were used for berry sampling.

2.2. Microclimate Assessment

The thermopluviometric characterization of the two seasons was assessed by bioclimatic indices calculated by the observation of an agrometeorological station belonging to the Regione Piemonte network. In order to characterize the four microenvironments in terms of radiation and thermal conditions, Photosynthetically Active Radiation (PAR) and air temperature inside the bunch zone were measured at intervals of 20′ each, from pea size stage to harvest as described in the previous research and according to an established protocol [31,33]. Then, the integral of daily amount of PAR (SPAR) and maximum daily temperatures (ST) and other cumulative thermal indices, such as Normalized Sum of the Degrees Celsius (SD) and Number of Hours (NH), were calculated over four periods corresponding to different phases of berry growth. The daily values of SPAR [MJ m^{-2} d^{-1}] were obtained by the cumulative sum of the hourly mean values of PAR irradiance [J m^{-2} s^{-1}] multiplied by 3600 s. NHs were the number of hours over the considered period in which the mean hourly maximum temperature (Tmax, [°C]) met three established ranges: Tmax ≥ 15 °C and Tmax < 25 °C (NH15 - 25); Tmax ≥25 °C and Tmax < 35 °C (NH25 - 35); Tmax ≥ 35 °C (NH ≥ 35). The same thresholds were used to calculate the SD indices by adding together the mean hourly maximum temperature (Tmax, [°C]) that were simultaneously higher than the minimum value of the threshold and lower than the maximum one, thus, SD15 - 25, SD25 - 35 and SD ≥ 35 respectively were assessed. To eliminate the influence of the period (from early summer to early autumn) in which the data were recorded, and of the different length of each period, the values of all variables were normalized by Equation (1):

$$\text{Normalized value} = (\text{VALUE} - \text{MEAN})/(\text{MAX} - \text{MIN}) \tag{1}$$

where VALUE was the value of the variable in a specific vineyard in a specific period; MEAN, MAX, and MIN, were, respectively, the mean, the maximum, and the minimum values of the variable calculated over all the vineyards in the considered period.

2.3. Berry Sampling

Four subsequent samplings of 400 berries were carried out randomly on each vineyard replicate, from about BBCH code75 (five–six weeks after bloom) until harvest [31]. In more detail:, in 2012, samplings were conducted on 12 July (about 24 days before veraison: dbV), 31 July (about 5 dbV), 27 August (about 22 days post veraison: dpV), 5 October (about 60 dpV); in 2013, 22 July (about 25 dbV), 21 August (about 5 dpV), 9 September (about 24 dpV), 15 October (about 60 dpV). For carotenoid analysis, three replicates of 50 g of berries were analyzed as described below.

2.4. Extraction and Determination of Carotenoids

2.4.1. Extraction from Grape Material

The procedure of carotenoid extraction was adapted from the method of Oliveira and others [21], as optimized by Crupi and collaborators [34]. Approximately 50 g of fresh berries, without seeds, added of 100 mg $Na_2S_2O_5$, were homogenized for 5 min in presence of magnesium carbonate basic. The homogenate was spiked with 200 µL of 183.2 mg/L of β-apo-8-carotenal (Fluka, Porto, Portugal, ref. 10,810) as internal standard, and diluted with 40 mL of water (Milli-Q, Millipore). Extraction was first carried out with 40 mL of ether/hexane (1:1, *v/v*), agitating for 30 min, then repeated twice with further 20 mL of ether/hexane. The upper layer was separated each time. The final extract was concentrated to dryness at 20 °C (Laborota 4001, Heidolph instruments) and resuspended in 1 mL of acetone/hexane (1:1, *v/v*) for High Performance-Liquid Chromatography (HPLC)/DAD determination. Each sample was injected in duplicate. Sample handling, homogenization, and extraction were carried out on ice under dim yellow light to minimize light-induced isomerization and oxidation of carotenoids.

2.4.2. High Performance-Liquid Chromatography (HPLC) Determinations

An Agilent Model 1200 quaternary solvent system equipped with quaternary pump solvent delivery and a UV-visible photodiode array detector was used. The absorption spectra were recorded at 447 nm and the sample injection volume was of 20 µL. The reversed stationary phase employed was a Lichrospher 100 RP C18 (5 µm) LichroCART (250 × 4 mm i.d.). Mobile phase was performed with solvent A: acetone/water (70:30 *v/v*), solvent B: acetone 100% (Sigma pure-grade), flow rate = 1 mL min^{-1}. The analytical gradient was: 0–20 min (from 100% to 0% of A), then from 20 to 30 min isocratic with 100% of B [25].

2.4.3. Identification and Quantification

The most relevant carotenoids were identified by comparison of UV-visible spectra with those of commercially available standards, β-carotene (Sigma 95%, synthetic,) (C-9750), lutein (Sigma 70%, from alfalfa) neoxanthin (0234.1) from CaroteNature GmbH (Erlenauweg 17, 3110 Münsingen, Switzerland), matching also different information such as position of absorption maxima (λ_{max}) and the degree of vibration fine structure (% III/II) (Table 1) [34].

Table 1. High Performance-Liquid Chromatography (HPLC)/DAD characteristics of carotenoids found in Nebbiolo grapes.

Compound	k'	λ_{max} (nm)	% (III/II) a
(9′ Z)-Neoxanthin	4.38	414; 436; 464	
(*all-E*)-Lutein	7.23	(422); 445; 472	42
β-Apo-8′-carotenal	9.28	460	
β-Carotene	11.82	(428); 452; 478	25

a % III/II is the ratio of the height of the longest-wavelength absorption peak, designated III, and that of the middle absorption peak, designated as II, taking the minimum between the 2 peaks as baseline.

Quantification of individual compounds was done by calibration curves using the respective standards for lutein and β-carotene with R^2 = 0.9997 and R^2 = 0.9991, respectively, whereas neoxanthin was expressed as lutein equivalent because of the unavailability of a fresh neoxanthin standard. The results were expressed in terms of concentration (mg kg^{-1} of berries) and content (µg $berry^{-1}$) to avoid an overestimation of the changes that may be the result of an altered berry surface area to volume ratio.

2.5. Statistical Analyses

Data were statistically analyzed by ANOVA (SPSS 15.0 for Windows, Chicago, USA and SAS 9.4 SAS Institute, Cary, USA), Tukey's test was used to assessed the differences as regards microclimatic variables and carotenoids both among treatments for each sampling date and between sampling dates for each treatment; the general effect of factors such as vineyard microenvironments (by treatment) and seasons (by year) and their possible interaction (year*treatment) were assessed too. Moreover, a 3-way-ANOVA, as regards carotenoid data of South treatments, was assessed, in order to evidence the vigor effect and interactions between year, sampling date and vigor level.

A preliminary ANOVA on normalized microclimatic variables (SD, NH, ST:SPAR) assessed the differences among sampling periods and the opportunity that these latter could be used as replicates for the comparison of vineyard microclimates and seasons. No differences among the periods emerged for none of the microclimatic variables; thus, the periods were used as replicates when microenvironments and seasons were compared. A Hierarchical Clustering Analysis (HCA) was also carried out using the method of centroide distance to evaluate the clustering of the vineyards based on the microclimatic conditions.

With the aim of identifying a model able to describe the impact of microclimate on berry composition, several Principal Component Analysis (PCA) were also performed on both microclimate (NH, SD, ST:SPAR) and carotenoids related variables including concentration (mg kg^{-1}), content (μg $berry^{-1}$), proportion (%) of lutein, β-carotene and neoxantin, lutein:β-carotene ratio, and sum of carotenoids (lutein+β-carotene+neoxanthin) as mg kg^{-1} of berries and μg $berry^{-1}$. The results reported here, refer to the data set of variables that explained the higher amount of the model variance. HCA and PCA were performed by SAS 9.4 (SAS Institute, Cary, USA).

3. Results

3.1. Seasonal Meteorological Trends

The two seasons presented some peculiarities from the meteorological point of view. In terms of annual values, the mean minimum and maximum temperatures were higher in 2012 than in 2013, as well as the average of the maximum monthly temperature that in 2012 exceeded the 2013 value by 2.7 °C. In the summer of 2012, the maximum temperatures exceeded 30 °C for 78 days, whereas in 2013 for 66. The hot condition of 2012 was also certified by the cumulative amount of Growing Degree Days (GDD) which exceeded the value of the 2013 by approximately 10%. Moreover, 2012 was drier than 2013, both in terms of rainfall amount and number of rainy days (> 1 mm). The differences between the years were also reflected on the growing period (Table 2). The warmest condition of 2012 affected the timing of the phenological phases that occurred earlier in 2012 than in 2013 [31]. In 2013, in fact, a delay of around 10 days for bud burst, a couple of weeks for bloom, a week for veraison and 10 days for commercial harvest was observed in comparison to 2012.

3.2. Bunch Microclimate

The thermal and radiative normalized indices calculated for each sampling date, were able to separate the two vineyard aspects when a cluster analysis was carried out, whereas clear separations between the levels of vigor, between seasons and among sampling dates were not evident (Figure 1 HCA). In more detail, ANOVA analysis, showed that differences among the four environments emerged for singular variables in both years (Table 3). When negative, the normalized indices indicated a negative difference compared to the average value calculated for all vineyards, and vice versa when positive. The higher the absolute values of the index, the higher were the differences. In 2012, the year with higher temperatures and lower rainfall SD15 - 25, NH15 - 25, and ST:SPAR were negative in the vineyards facing south, therefore, lower than the average calculated on all the vineyards, and they were significantly different from those of the vineyards facing west. In addition the vigor of the plants affected NH15 - 25 and ST:SPAR indices of the south facing vineyards showing both a lower value in the less vigorous condition (V−). No differences among vineyards emerged for the indices referring to the other temperature ranges.

Table 2. Meteorological characterization of the two seasons (the values are calculated both for the entire year and for the grapevine vegetative period: April–October). Data were registered by Serralunga Boscareto's meteorological station (Agrometeorological Network, Regione Piemonte).

	Tmax Mean (°C)	Tmin Mean (°C)	Average of the Monthly Maximum Values of Tmax (°C)	Total Rainfall (mm)	Growing Degree Days (Base 10 °C)	Days with T >30 °C (Number)	Rainy Days[1] (Number)
Annual Values							
2012	20.4	8.9	28.3	722	2329	78	63
2013	19.4	8.6	25.6	971	2119	66	83
Values of April–October Period							
2012	26.0	13.8	32.2	491	2147	78	36
2013	25.6	13.4	31.8	563	2046	66	49

[1] Days with precipitation over 1 mm.

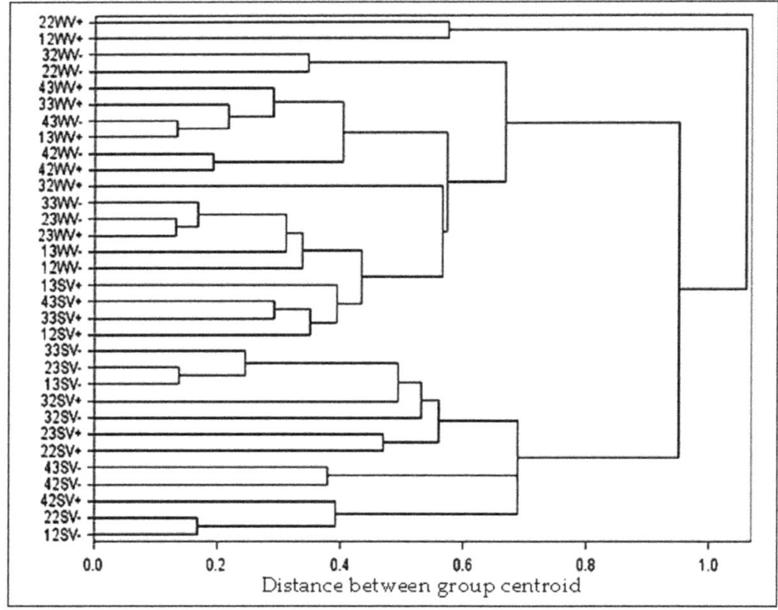

Figure 1. Dendrogram of hierarchical clustering analysis of the four microenvironment (SV−, SV+, WV−, and WV+) obtained by analyzing the meteo-climatic indices reported in Table 2. (1, 2, 3, and 4, as the first digit in the treatment acronym, correspond to the phenological period; 2, 3, as the second digit, correspond to the season 2012 and 2013, respectively; W and S represent the West and South vineyard aspect, respectively; V+ and V− indicate a higher or lower plot vigor, respectively).

Table 3. Normalized Sum of the Degrees Celsius (SDs), number of hours (NHs) related to the temperature ranges (15– 25; 25– 35; ≥35) and ST:SPAR; ST=summation of the maximum daily temperatures; SPAR=summation of the daily integrals of PAR.

	Treatment	SDs Celsius (°C)			NHs Number of Hours NHT Number of Hours NHT			ST:SPAR
		15–25	25–35	≥35	15–25	25–35	≥35	
2012	SouthV−	−0.54 a	0.18 a	0.12 a	−1.20 a	0.50 a	0.58 a	−0.56 a
	SouthV+	−0.10 ab	−0.06 a	−0.05 a	−0.37 b	−0.02 a	0.20 a	−0.18 b
	WestV−	0.40 b	0.22 a	−0.43 a	0.78 c	0.45 a	−1.05 a	0.29 c
	WestV+	0.23 b	−0.34 a	0.36 a	0.80 c	−0.92 a	0.27 a	0.44 c
		***	ns	ns	***	ns	ns	***
2013	SouthV−	−0.59 a	−0.23 a	0.66 c	−0.59 a	−0.34 a	0.63 c	−0.59 b
	SouthV+	0.03 b	0.28 a	−0.07 b	−0.08 b	0.12 a	−0.01 b	−0.22 b
	WestV−	0.23 b	0.22 a	−0.34 a	0.29 b	0.28 a	−0.37 a	0.40 c
	WestV+	0.32 b	−0.27 a	−0.24 ab	0.37 b	−0.06 a	−0.26 a	0.41 c
		***	ns	***	***	ns	***	***
Treatment		***	ns	***	***	ns	***	***
Year		ns	ns	ns	ns	ns	ns	ns
Year*Treatment		ns	ns	***	***	ns	ns	ns

For the same year, means fallowed by different letters are significant different for $p < 0.05$. ns—not significant; * $p < 0.05$, *** $p < 0.001$ indicate the significance of differences among treatments.

In 2013, the coolest and wettest year, the differences emerged in the first year were confirmed. Furthermore, SD ≥ 35 and NH ≥ 35, were higher in the vineyards facing south and, in particular, in the less vigorous ones. On the contrary, in the vineyards facing west, the influence of vigor did not emerge. Therefore, differences among vineyards were evident for the indices relating to the lowest and highest temperature ranges whereas for the intermediate range (25 - 35) both SD and NH were similar in all the vineyards. For none of the indices, differences between the years emerged while the year * vineyard interaction was significant only for SD ≥ 35 and NH15 - 25.

3.3. Evolution of Carotenoids in Nebbiolo Grapes

The compounds that mostly contributed to the total amount of carotenoids in Nebbiolo grapes were lutein and β-carotene. The evolution of each compound during ripening, both as µg berry^{-1} (Figure 2a, Figure 3a, and Figure 4a) and as mg kg^{-1} of fresh berries (Figure 2b, Figure 3b, and Figure 4b), is shown for both years 2012 and 2013 (Table 4).

● Lutein

In 2012, at the first sampling date, at 24 days before veraison (dbV) similar amounts of lutein were detected in green berries from both west and south exposed vineyards (Figure 2a). Lutein concentration (mg kg^{-1} of berries) and content (µg berry^{-1}) remained then constant until the beginning of veraison (5 dpV, second sampling) (Figure 2a,b and Table 4). In 2013, at the first sampling (25 dbV), the lutein concentration (mg kg^{-1}) was significantly higher in SouthV+ when compared to the other treatments and to the previous year. Afterwards, at the second sampling (5 dpV), lutein content decreased significantly only in south-exposed vineyards (Figure 2a,b, Table 4).

Table 4. Seasonal changes of the carotenoid concentration and of the ratio Lutein: β-carotene in Nebbiolo grapes during 2012 and 2013.

Year	Treatment	2012					2013							
Sampling Date		12 Jun	31 Jul	27 Aug	5 Oct	sig. §	22 Jul	21 Aug	9 Sep	15 Oct	sig. §	sig. §		
Phenological Phase		24dbV	5dbV	22dpV	60dpV	date	25dbV	5dpV	24dpV	60dpV	date	T	Y	Y*T
Lutein mg kg^{-1}	SouthV−	4.84b	4.52b	4.37b	1.65a	**	6.19b	4.09a	3.53a	2.52a	*			
	SouthV+	5.29b	4.73b	5.60b	2.68a	*	9.69b	4.13a	5.64a	3.70a	*	***	***	ns
	WestV−	-	-	-	-		5.39a	5.60a	5.59a	3.95a	ns			
	WestV+	4.42ab	3.76a	5.76b	3.75a	**	5.41ab	5.09ab	6.55b	4.25a	*			
Sig. among treat.		ns	ns	ns	*		**	ns	*	**				
Lutein µg berry^{-1}	SouthV−	3.56a	3.81a	6.80 b	2.78a	**	5.35a	4.76a	5.70a	4.40a	ns			
	SouthV+	4.21a	4.18a	9.07b	4.66a	**	6.90b	4.89a	9.30c	7.04b	**	***	***	ns
	WestV−	-	-	-	-		4.68a	6.58ab	9.51b	7.41ab	**			
	WestV+	3.94a	3.76a	10.1c	6.75b	**	4.80a	5.84a	11.3b	7.91ab	*			
Sig. among treat.		ns	ns	*	*		ns	ns	*	**				
β-carotene mg kg^{-1}	SouthV−	8.11b	5.37ab	3.60a	2.53a	**	10.3b	2.32a	3.08a	3.18a	**			
	SouthV+	9.57b	6.37ab	4.46a	3.64a	**	13.7c	2.49a	4.43b	3.51ab	**	***	ns	ns
	WestV−						10.1c	5.55b	4.17ab	3.48a	**			
	WestV+	7.47b	5.91ab	3.92a	4.07a	**	10.4b	5.64a	5.18a	4.22a	**			
Sig. among T		ns	ns	ns	*		ns	***	ns	*				
β-carotene µg berry^{-1}	SouthV−	5.97a	4.53a	5.61a	4.24a	ns	8.91b	2.70a	4.97a	5.56a	**			
	SouthV+	7.61a	5.63a	7.21a	6.26a	ns	11.4c	2.95a	7.31b	6.70ab	**	***	**	ns
	estV−	-	-	-	-		8.70a	6.52a	7.08a	6.52a	ns			
	WestV+	6.66a	5.91a	6.91a	7.33a	ns	9.24a	6.47a	8.98a	7.88a	ns			
Sig. among T		ns	ns	ns	*		ns	**	*	*				
Neoxantine mg kg^{-1}	SouthV−	nd	0.14a	0.41b	0.11a	**	0.06a	0.24a	0.15a	0.07a	ns			
	SouthV+	nd	0.12a	0.55b	0.17a	**	0.18ab	0.25b	0.26c	0.10a	**	*	ns	ns
	WestV−	-	-	-	-		0.08a	0.15ab	0.34c	0.23b	**			
	WestV+	nd	0.03a	0.51c	0.25b	**	0.06a	0.14ab	0.41c	0.18b	**			
Sig. among T		–	**	ns	*		ns	ns	ns	**				
Neoxantine µg berry^{-1}	SouthV−	nd	0.12a	0.64b	0.18a	**	0.08a	0.28ab	0.25b	0.12ab	*			
	SouthV+	nd	0.11a	0.89b	0.30a	**	0.15a	0.30a	0.43b	0.19a	**	**	ns	***
	WestV−	-	-	-	-		0.07a	0.17a	0.58b	0.44b	**			
	WestV+	nd	0.03a	0.91c	0.46b	**	0.06a	0.16a	0.70c	0.34b	**			
Sig. among T		–	*	ns	*		ns	ns	ns	**				
Lutein:β-carotene	SouthV−	0.60a	0.84a	1.21b	0.65a	**	0.60a	1.75b	1.15ab	0.79a	***			
	SouthV+	0.55a	0.75a	1.25b	0.73a	**	0.72a	1.65b	1.28ab	1.05ab	**	**	***	***
	WestV−	-	-	-	-		0.52a	0.91ab	1.27b	1.01a	*			
	WestV+	0.59a	0.63a	1.49c	0.93b	**	0.54a	1.01b	1.34b	1.13b	*			
Sig. among T		*	***	ns	*		ns	***	*	***				
[1] Sum of carotenoids mg kg^{-1} of berries	SouthV−	13.0c	10.0bc	8.38b	4.30a	**	16.6b	6.6a	6.84a	5.81a	**			
	SouthV+	14.9b	11.2ab	10.6ab	6.49a	*	23.5c	6.9a	10.5b	7.31a	***	***	**	ns
	WestV−	-	-	-	-		15.5b	10.9a	10.1a	7.67a	*			
	WestV+	11.9a	9.7a	10.2a	8.07a	**	15.9b	11.3ab	12.1ab	8.65a	*			
Signif. among T		ns	ns	ns	*		**	**	*	*				
Sum of carotenoids µg berry^{-1}	SouthV−	9.53ab	8.46ab	13.1b	7.21a	**	14.3b	7.73a	11.0ab	10.1ab	*			
	SouthV+	11.9ab	9.91a	17.2b	11.2ab	**	18.5c	8.14a	17.3bc	13.9b	**	***	***	ns
	WestV−	-	-	-	-		13.4a	12.5a	17.2a	14.4a	ns			
	WestV+	10.6ab	9.67a	18.0c	14.5bc	**	14.1a	13.3a	21.0a	16.1a	ns			
Sig. among T		ns	ns	ns	*		ns	*	*	**				

For the same line and year means followed by different letters indicate significant differences among sampling dates for 2012 and 2013 respectively. *, **, *** indicate, respectively, significant differences for $p < 0.05$, $p < 0.01$, and $p < 0.001$. nd—not detectable, ns—not significant, sign §—statistical differences by date, by T—Treatment; by Y—year; T*Y—interactions between Treatment and Year; [1]—Lutein+β-carotene+neoxanthin.

Concurrently to the increase of the berry weight and sugar concentration [31], a significant increase of lutein content per berry was noticed in both years reaching a peak at about 4 weeks after veraison, with the exception of SouthV− in 2013. This increase of lutein content was proportional to the vine vigor, thus more important for the most vigorous vines. After veraison, in SouthV− a minor increase was observed in 2012 and a constant trend in 2013, thus, at the time of the peak, significant differences were noticed between WestV+ and SouthV− in both seasons. At the final stage of ripening, a significant lutein degradation was observed for all treatment in 2012; significant differences between SouthV− (1.66 mg kg^{-1} of berries or 2.78 µg berry^{-1}) and WestV+ parcels (3.75 mg kg^{-1} of berries or 6.78 µg berry^{-1}) were found at harvest (Figure 2a,b and Table 4). The concentration decline observed in 2013 was less remarkable for all vineyards compared to 2012, resulting significant only for the most vigorous plots, WestV+ (mg kg^{-1} of berries) and SouthV+ (µg berry^{-1}). The content of lutein in the less vigorous SouthV− at harvest was significantly lower (4.4 µg berry^{-1} or 2.52 mg kg^{-1} of berries) than in the other vineyards. Thus, the warmest vineyard had the lowest lutein concentration at harvest in both years. Lower amounts of lutein were detected at harvest in 2012 for all treatments, but significantly lower only for SouthV−, when compared to 2013 (average at harvest, 2.69 mg kg^{-1} and 4.73 µg berry^{-1} in 2012; 3.61 mg kg^{-1} and 6.7 µg berry^{-1} in 2013). The interaction year*treatment was not significant regardless of the unit (Table 4).

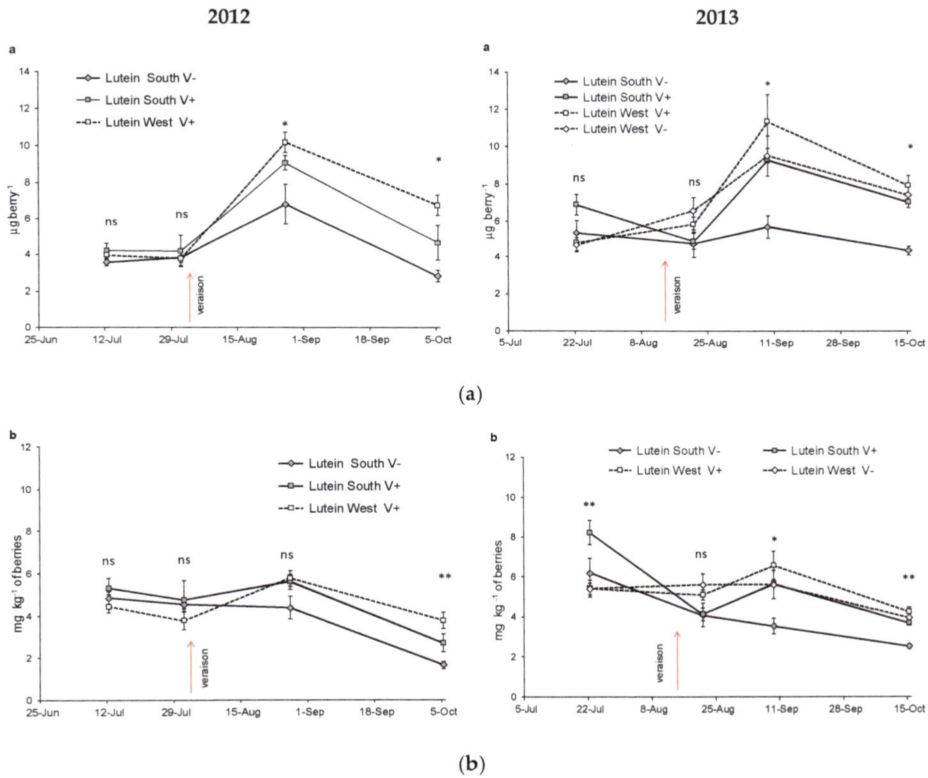

Figure 2. Seasonal changes of lutein as µg berry^{-1} (**a**) and mg kg^{-1} of berries, (**b**) in Nebbiolo grapes in 2012 (on the left) and 2013 (on the right), as a function of vine vigor and vineyard aspect. Averages ± standard error (*n* = 3). * *p* < 0.05, ** *p* < 0.01 indicate the significance of differences among treatments.

- β-Carotene

In 2012, the β-carotene concentrations (mg kg^{-1} of berries) decreased during the season with no significant differences among treatments at any sampling point except at harvest, when significantly higher amounts were noticed in the WestV+ (4.07 mg kg^{-1} or 7.33 µg berry^{-1}) than in the SouthV− grapes (2.54 mg kg^{-1} or 4.24 µg berry^{-1}). In 2013, the β-carotene decline between the first and second sampling was significant for south-exposed vineyards while the post veraison increase of β-carotene content in 2013, was significant only for the SouthV+ grapes (Figure 3a,b, Table 4). β-carotene content (µg berry^{-1}) and concentration (mg kg^{-1} of berries) were significantly lower in SouthV− than in

WestV+ both at the third sampling and at harvest (3.2 mg kg^{-1} or 5.6 µg berry^{-1} for SouthV– and 4.2 mg kg^{-1} or 7.9 µg berry^{-1} for WestV+). Moreover, differences between the two years at harvest, were not significant when comparing the values as mg kg^{-1} (averagely 3.41 versus 3.60, in 2012 and 2013, respectively) but they were significant when comparing the values as µg berry^{-1} (5.94 versus 6.7, in 2012 and 2013, respectively). The interaction year*treatment was not significant regardless the unit (Table 4).

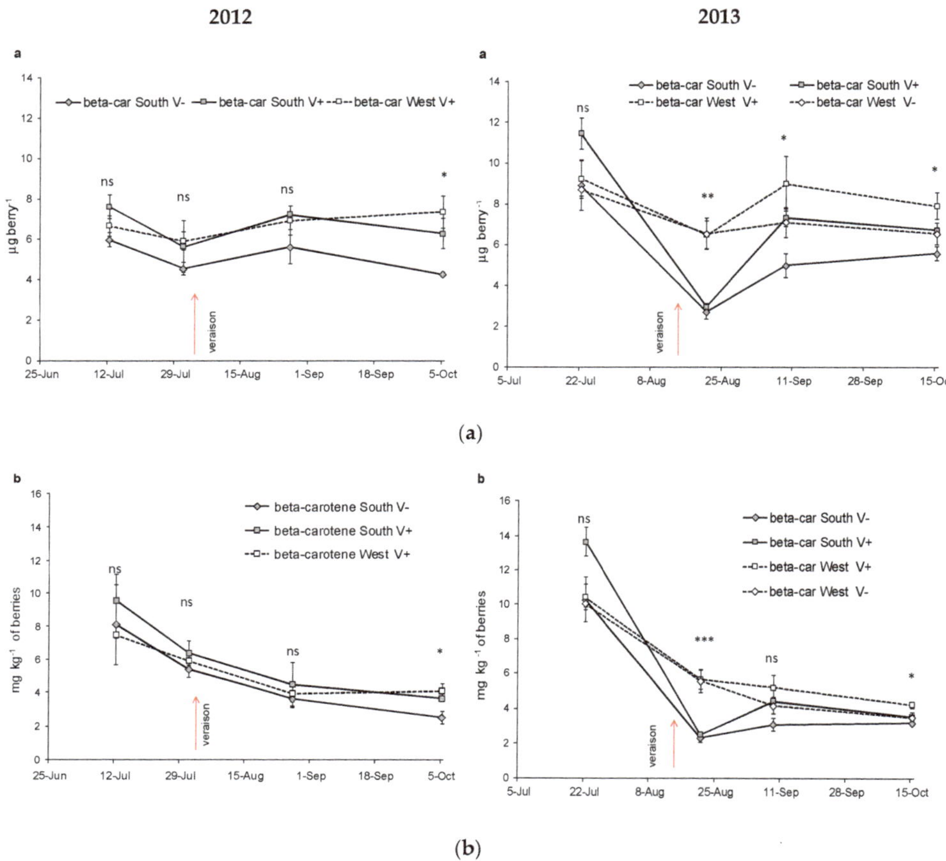

Figure 3. Seasonal changes of β-carotene as µg berry^{-1} (**a**) and mg kg^{-1} of berries (**b**) in Nebbiolo grapes in 2012 (on the left) and 2013 (on the right), as a function of vine vigor and vineyard aspect. Averages ± standard error (n = 3). * $p < 0.05$, ** $p < 0.01$ *** $p < 0.001$ indicate the significance of differences among treatments. b.

- Neoxanthin

In the green berries, undetectable amounts (2012) or traces (2013) of neoxanthin were observed for all treatments. The important increase after veraison, observed for the most vigorous plots was more remarkable in 2012 than in 2013, and was then followed by a significant degradation of this compound until harvest. In 2012, as for lutein and β-carotene, significant differences were observed at harvest between SouthV– (0.11 mg kg^{-1} of berries or 0.18 µg berry^{-1}) and WestV+ grapes (0.25 mg kg^{-1} of berries or 0.46 µg berry^{-1}). Similarly to lutein, a peak of concentration was achieved 4 weeks after veraison but, differently from lutein and β-carotene, was higher in 2012 (averagely 0.49 µg kg^{-1} and 0.81 µg berry^{-1}) than in 2013 (averagely 0.30 µg kg^{-1} and 0.49 µg berry^{-1}) (Figure 4).

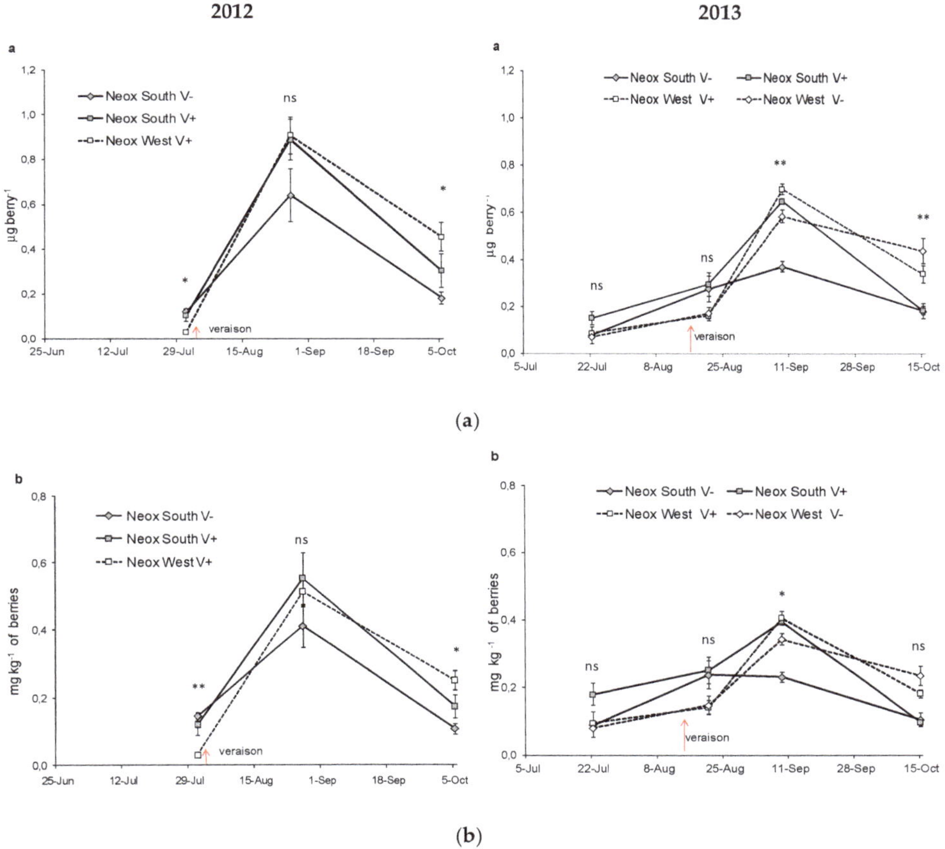

Figure 4. Seasonal changes of neoxanthine as µg berry^{-1} (**a**) and mg kg^{-1} of berries (**b**) in Nebbiolo grapes in 2012 (on the left) and 2013 (on the right), as a function of vine vigor and vineyard aspect. Averages ± standard error (n = 3). * $p < 0.05$, ** $p < 0.01$ indicate the significance of differences among treatments.

In both years, this peak was followed by a significant degradation of neoxanthin until harvest period in SouthV+, WestV−, and WestV+ (Figure 3a). Moreover, in both years, a high degradation rate was highlighted especially for the most vigorous SouthV+ that influenced neoxanthin amount at harvest time. In SouthV−, instead and similarly to lutein, neoxanthin levels followed a more flattened evolution showing a minor peak mostly in 2013. In both years after veraison, SouthV− attained the lowest values compared to the other treatments especially when µg berry^{-1} were considered. The vine vigor (V−, V+) did not affect the neoxanthin content at harvest. Differences between the years were not significant; the interaction year*treatment was significant when values were expressed as µg berry^{-1} (Figure 4, Table 4).

• Lutein: β-Carotene Ratio and Sum of Carotenoids

The ratio lutein: β-carotene increased between green phase (first sampling) and complete veraison (third sampling), whereas a decrease was evident during the weeks preceding commercial harvest. In 2012, the ratio was in favor of β-carotene whereas in 2013 the ratio was often in favor of lutein. Comparing both years, the ratio varied at harvest from 0.65 to 0.79 in the warm and more exposed plots (SouthV−) and from 0.93 to 1.13 in the cooler and more vigorous WestV+. In average, the difference between years was significant as well the interaction years*treatments (Table 4). In general, the sum of the concentration (mg kg^{-1}) of the three considered carotenoids, decreased from the first to the last sampling, whereas, when expressed as µg berry^{-1}, a peak was observed at the third sampling in both years. However, at harvest in 2012 and at all sampling dates in 2013, the sum of the carotenoids (concentration and content) was lower in SouthV−, than in WestV+.

Summarizing, the season significantly influenced all the variables except the neoxanthine and β-carotene concentration. Regardless the season, the treatment significantly influenced the content of all compounds. Nevertheless, the interaction year*treatment was significant only for neoxantine (µg berry^{-1}) and lutein: β-carotene ratio. Focusing on the results of the 3-way ANOVA, carried out on the south facing vineyards, significant differences between the two levels of vine vigor emerged for all compounds, but neither the interaction year*vigor nor year*vigor*sampling date were significant, regardless the unit of measurement of carotenoids (Supplementary Table S1).

Performing a PCA on a data matrix including microclimatic indices and berry carotenoids an effective distribution of the vineyards along the first component (PC1), and of the sampling dates along the second component (PC2), emerged (Figure 5).

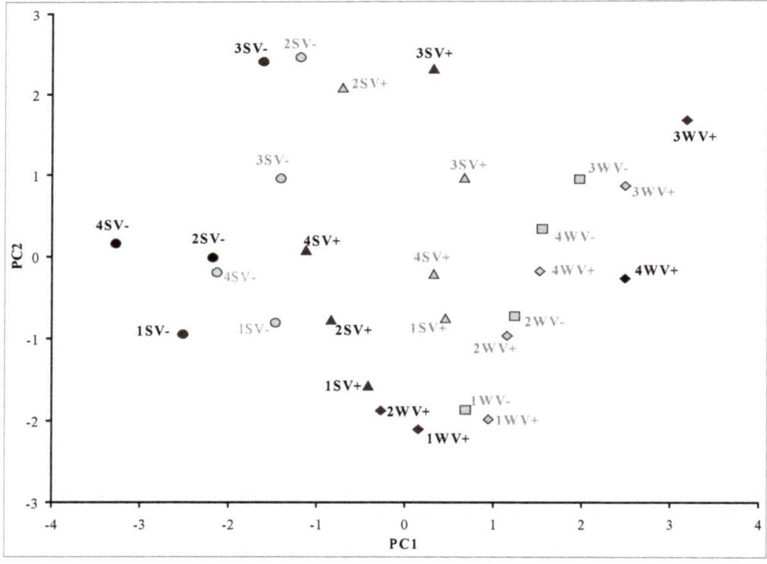

Figure 5. Score plot of the observations in the Cartesian coordinate system identified by Principal Component Analysis: analysis including variables related to carotenoids and microclimatic indexes (the variables included in the analysis are reported in Table 4). 1, 2, 3, and 4 refer to the data sampling; S—South, W—West, V+—high vigor, V−—low vigor. Symbols in black color—2012; symbols in grey color—2013; ●—SouthV−, ▲—SouthV+, ■—WestV−, and ♦—WestV+.

Prin1 and Prin2 explained, respectively, 39% and 27% of the total variance. The case distribution along Prin1 was well represented by a linear combination of the variables NH15 - 25, ST:SPAR and sum of carotenoids (positively correlated with Prin1). The case distribution along Prin2 was well represented by a linear combination of the variables Neoxanthine (%) and lutein: β-carotene ratio (positively correlated with Prin2) (Table 5).

Table 5. Principal Component Analysis (PCA): Percentage of variance explained by the three principal components (Prin1, Prin2, and Prin3), eigenvalues and loadings indicating the correlation between the variables and the three principal components.

	Prin1	Prin2	Prin3
Explained variance (%)	39	27	20
Eigenvalues	2.7	1.9	1.4
Loadings of the variables			
Neoxanthin (%)	0.15	**0.67**	0.02
Sum of carotenoids (μg berry^{-1})	**0.46**	0.09	−0.12
Lutein: β-carotene	0.18	**0.65**	0.04
NH15–25	**0.53**	−0.20	0.23
NH25 -35	−0.13	−0.02	**−0.79**
NH35	−0.41	0.05	**0.54**
ST:SPAR	**0.51**	−0.26	0.09

4. Discussion

The two seasons of the study were different from a weather point of view. In fact, 2012 was generally warmer and drier than 2013 (Table 1). The conditions of 2013 enhanced vine vigor avoided summer water stress and delayed the timing of all phenological phases. Nevertheless, dry and warm condition from mid-September to mid-October prolonged vine metabolic activity, and allowed it to reach an optimal level of berry ripening.

The thermal and radiative microclimate indices allowed separating south facing vineyards from the west facing ones, but neither the sampling dates, nor the vigor levels (Figure 1). In general, the southern vineyards were the hottest even if, as evidenced by the significant interaction year*treatment for SD ≥ 35 and NH15 - 25, the differences between the south and west vineyards were amplified in the cooler and wetter season (2013) whereas they were mitigated in the warmer season (2012). ST:SPAR was smaller in the southern than in the western vineyards in both years. This index is an expression of two synergistic effects. Firstly, in hilly conditions such as those of the study, the southern aspect intercepted higher amount of solar radiation (higher values of SPAR). Secondly, the west side of the hill registered the maximum daily temperature (thus, the high value of ST) when, in the afternoon, the Sun's rays were perpendicular to the slope. This index showed a good potential in characterizing the microclimates and it contributed to their separation when inserted into the dataset analyzed by PCA.

The PCA analysis conducted on both microclimatic indices and carotenoid compounds, effectively separated the west facing vineyards from the south facing ones and, in the latter case, also the two levels of vigor (Figure 5). The vineyards separation along Prin1 was determined by the sum of the carotenoids and by both the microclimatic indices NH15 - 25 and ST:SPAR, whose values were greater in the west facing plots. All these variables were positively correlated with Prin1 (Table 5), thus, the overall amount of the carotenoids was favored by cooler conditions. The differences between the vintages and between the levels of vigor were evident in the warmer vineyards where the intercepted radiation was greater (South) but not in the fresher ones (West) whose conditions were evidently less favorable for the degradation of carotenoids (Figure 5). This observation led to think that the negative effect of higher temperatures prevails over the positive effect of a higher interception of radiation on grape carotenoid concentration at harvest. In the first part of this study [31], higher amount of C13 norisoprenoids was found in the sunniest year and in the more sunlit vineyards. The decrease in the concentration of those compounds in the warmer SouthV− plot in the last ripening stage was likely attributable to the effect of high temperatures. This decrease was more evident for some compounds and less for others, thus microclimate had an effect on both concentration and profile of C13-norisoprenoids in Nebbiolo grapes [31] confirming, at least in part, results already issued [14].

The higher carotenoid content found in Nebbiolo grapes when compared to other varieties [35–38], agrees with the levels found so far for cv. Nebbiolo [28]. A peak of lutein concentration, that was proportional to the vine vigor, was noticed after veraison in both seasons. A similar peak was reported for Touriga Franca [21], Nebbiolo, and Barbera [28] and, more recently, for Merlot [36]. Therefore, it is possible that cultivar differences exist in the timing of lutein synthesis and degradation thus, a probable delay in VvCCD1 expression can be hypothesize for these varieties [8]. In the present study, the lutein peak after veraison was more evident in the most vigorous plants. This result may be attributable to the high vigor which often reflects a greater synthesis of chlorophyll so, probably, of carotenoids and/or to the lower degradation of this compound under the cooler conditions of these plants. Furthermore, a higher content of lutein-5,6-epoxide (not quantified in this study) in shaded berries prior the veraison can be presumed; the accumulation of this compound has been shown to be a plant early response to shade conditions [8,15]. Therefore, a higher transformation of this compound in lutein may occur after veraison. In both years, the peak of lutein after veraison was less evident in SouthV−; high temperatures and a higher exposure to sunlight probably promoted the carotenoid degradation in post veraison, as proposed in literature [39]. Lutein is also reported to be more efficient than violaxanthin in preventing ROS formation, thus, it could be further used by grapevines for photoprotection under stress conditions [40]. This could also explain the lower lutein concentration and its higher rate of degradation under more stressing condition, such as in south facing vineyard and in the warmer periods of the season, where high radiation and temperature (>35 °C) were achieved for many hours a day in both years [31]. As regards the other treatments, the west-exposed and the more vigorous plots maintained the highest amounts at harvest. Our findings agree with previous research reporting that grapes grown in shaded conditions [21,41] had higher carotenoid levels. On the contrary, under high UV-B levels, lower concentrations of total carotenoids were found [27].

In 2012, constant levels for β-carotene were observed during ripening in all experimental plots when results were expressed in μg berry^{-1}, while a significantly lower amount was registered in SouthV− in both years at harvest. In 2013, instead, higher contents of β-carotene than in 2012 were highlighted in green berries (three weeks before veraison) and for the south facing vineyards a notable decrease was registered thereafter during veraison. The decline of β-carotene during that period was already observed for Nebbiolo and Barbera grapes and was attributed to the high temperature [28]. According to other studies, instead, β-carotene content in grapes, shows an increase from preveraison to veraison, and a decline thereafter until harvest [23,25,36]. In our study, the condition of west facing vineyards, less sunlit and lower temperature than in south ones, likely led to a lower degradation of this compound allowing the maintenance of a constant content (μg berry^{-1}) in both years. Other research reported also a greater impact of microclimatic variations on lutein more than on β-carotene [42]. Our results showed a different impact of microclimate on β-carotene if compared to the two xanthophylls since high temperatures of south vineyards in the second year led to a higher degradation of β-carotene during veraison more than at the final stages of ripening.

The warmer conditions registered 4 weeks after veraison in the second year of the study, probably penalized also the concentration (μg kg^{-1}) and content (μg berry^{-1}) of the neoxanthin that resulted particularly lower at the supposed peak moment, in the south-facing plots. The behavior of this compound was similar to lutein.

The lutein: β-carotene ratio is an indicator of flux to the a- and b-branches, respectively, of the carotenoid metabolic pathway [15]. In literature is reported an influence of the variety on this ratio; in some cultivars indeed (i.e., Syrah, Sauvignon, Pinot Noir, and Merlot), the lutein level was almost twice than that of β-carotene. In Chardonnay and Carignan, the concentration of the two carotenoids was very similar while a higher level of β-carotene was found in Grenache, Gamay, and Sauvignon blanc [15,25,26,38]. In addition, the growing region and topographic features of the site [34,38], as well as cultural practices, such as leaf removal [15], may also affect this ratio. The lutein: β-carotene ratio in

Nebbiolo grapes, varied during berry development also depending on season and vineyard aspect (Table 4). Lutein prevailed on β-carotene after veraison. However, this did not happen, neither in the early stages nor at harvest, nevertheless, in both years, the ratio in the later phases was lower in the warmest nor more exposed plot in accordance with previous research [15,19].

Recent research on the effect of bunch zone leaf removal on Sauvignon blanc [15] concerned mostly the light effect since irrelevant temperature differences within the bunches were registered. Under these conditions, the concentration of the major carotenoids decreased during berry development, following the behavior of chlorophylls, whereas specific xanthophylls, such as lutein, resulted more abundant during the early stages of berry development in berries more exposed to sunlight. In addition, as already reported, the intensity of solar radiation exerts a great role on the degradation of grape carotenoids [26]. Nevertheless, another study highlighted that the carotenoid concentration from veraison to harvest was positively correlated with temperature but less correlated with both rainfall and radiation [19]. According to our study, individual carotenoids respond in a different way depending on the microclimatic conditions of each specific period during ripening. Nevertheless, differently from the previous studies, a major synthesis of lutein occurred after veraison in the vigorous plots where both radiation and temperature were lower whereas in the warmer and more exposed plots (SouthV−), lutein accumulation and final content were penalized (Figure 2). β-carotene content was lower in the warmer period and in the more sunlit south plots with respect to the cooler and more shaded west plots, as well. Actually, the differences emerged between the growing seasons and between vineyard aspects highlighted the influence of the temperature on the rate of synthesis and/or degradation of β-carotene, (Figure 3, Table 4). In any case, the positive effect of radiation did not clearly emerge in our study since the highest peaks were recorded in the less sunlit vineyards (Figure 2).

According to the literature, the amount of berry carotenoids at harvest seemed to be more dependent on the condition of the earlier developmental stage that impact on their synthesis, rather than on the condition of the final phases that impacts their degradation [26]. As a result of the current study, the lower temperature of the early phases of 2013, favored the amounts of carotenoids in the green berries, even in conditions of lower radiation [31]. The particularly warmer condition of the period after veraison in the less vigorous south plots, promoted a higher degradation even with different rate depending on the compound. Nevertheless, a higher total content (μg berry^{-1}) of carotenoids was averagely measured at harvest in 2013 than in 2012, confirming the importance of the accumulation phase on the final content at harvest [26].

5. Conclusions

This study confirms most results of previous studies and illustrates the effect of vineyard aspect and bunch microclimate on both synthesis and degradation of the most relevant grape carotenoids during berry development in Nebbiolo grapes. Normalized microclimate indices appeared useful to characterize growing seasons and the microenvironments and to explain the compositional differences between the examined environments.

Lutein and neoxanthin responded in a similar way to environment variability having a similar peak after veraison in both years, whereas a variety effect can be presumed as regards lutein trend during ripening. A different trend was observed for β-carotene content depending on season and vineyard aspect. Generally, the warmer conditions of the most sunlit south facing vineyards led to low amounts of all compounds at harvest. On the other hand, less warm conditions, like those of west-exposed vineyards or more vigorous vines, likely favored the synthesis and/or induced a lower degradation of carotenoid compounds. Nevertheless, carotenoids seemed to respond to microclimate variability differently depending on the compound. In addition, the amount of radiation accumulated in specific periods, and mainly the prolonged high temperatures during the last stages of grape ripening, determined the evolution of carotenoids during season and their profile and quantity at harvest. Furthermore, a repeatability of these results can be expected since the relationships observed among the environments were maintained in both years and despite the overall differences between them from a meteorological point of view.

Our results highlight also that in earlier vintages, driven by the climate warming, the grapes ripen in a warmer period when high temperature determine a higher degradation influencing considerably both berry carotenoid profile and concentration. Nowadays, winegrowers are called to face such warmer climatic conditions, therefore, our study add further knowledge to target vineyard cultural strategies in order to modulate aroma potential of grapes.

Supplementary Materials: The following are available online at http://www.mdpi.com/2076-3417/10/11/3846/s1, Table S1: Supplementary Table S1. Results of the 3-way ANOVA, carried out on the carotenoid content (µg berry-1) and concentration (mg kg-1) of the South facing vineyard considering the two levels of vigour (V+ and V-), the two seasons and the four sampling dates as factors of variability.

Author Contributions: Conceptualization, S.G.; Formal analysis, A.A., S.C., E.M. and S.G.; Funding acquisition, S.G.; Investigation, A.A. and E.M.; Methodology, A.A., S.C. and S.G.; Resources, M.P.; Supervision, S.G.; Validation, M.P. and S.C.; Writing–original draft, A.A.; Writing–review & editing, A.A., M.P., S.C., A.F. and S.G. All authors have read and agreed to the published version of the manuscript.

Funding: This research received no external funding

Acknowledgments: The authors would like to thank Fondazione Dalmasso and the Azienda Agricola G.D. VAJRA in Barolo, for supporting this study.

Conflicts of Interest: The authors declare no conflict of interest.

References

1. Hirschberg, J. Carotenoid biosynthesis in flowering plants. *Curr. Opin. Plant Biol.* **2001**, *4*, 210–218. [CrossRef]
2. Winterhalter, P.; Ebeler, S.E. Carotenoid cleavage products: An Introduction. In *Carotenoid Cleavage Products*; American Chemical Society: Washington, DC, USA, 2013; Volume 1134, pp. 3–9. ISBN 0-8412-2778-0.
3. Mendes-Pinto, M.M. Carotenoid breakdown products the—norisoprenoids—in wine aroma. *Arch. Biochem. Biophys.* **2009**, *483*, 236–245. [CrossRef]
4. Winterhalter, P.; Rouseff, R.L. *Carotenoid-Derived Aroma Compounds*; American Chemical Society: Washington, DC, USA, 2002; Volume 802.
5. Skouroumounis, G.K.; Sefton, M.A. Acid-catalyzed hydrolysis of alcohols and their β-d-Glucopyranosides. *J. Agric. Food Chem.* **2000**, *48*, 2033–2039. [CrossRef] [PubMed]
6. Mathieu, S.; Terrier, N.; Procureur, J.; Bigey, F.; Günata, Z. A carotenoid cleavage dioxygenase from Vitis vinifera L.: Functional characterization and expression during grape berry development in relation to C13-norisoprenoid accumulation. *J. Exp. Bot.* **2005**, *56*, 2721–2731. [CrossRef] [PubMed]
7. Walter, M.H.; Strack, D. Carotenoids and their cleavage products: Biosynthesis and functions. *Nat. Prod. Rep.* **2011**, *28*, 663–692. [CrossRef] [PubMed]
8. Young, P.R.; Lashbrooke, J.G.; Alexandersson, E.; Jacobson, D.; Moser, C.; Velasco, R.; Vivier, M.A. The genes and enzymes of the carotenoid metabolic pathway in Vitis vinifera L. *BMC Genom.* **2012**, *13*, 243. [CrossRef]
9. Baumes, R. Wine Aroma Precursors. In *Wine Chemistry and Biochemistry*; Moreno-Arribas, M.V., Polo, M.C., Eds.; Springer: New York, NY, USA, 2009; pp. 251–274. ISBN 978-0-387-74116-1.
10. Silva Ferreira, A.C.; Monteiro, J.; Oliveira, C.; Guedes de Pinho, P. Study of major aromatic compounds in port wines from carotenoid degradation. *Food Chem.* **2008**, *110*, 83–87. [CrossRef]
11. Guedes de Pinho, P.; Silva Ferreira, A.C.; Mendes Pinto, M.; Benitez, J.G.; Hogg, T.A. Determination of carotenoid profiles in grapes, musts, and fortified wines from Douro varieties of Vitis vinifera. *J. Agric. Food Chem.* **2001**, *49*, 5484–5488. [CrossRef]
12. Goodwin, T.W. Metabolism, nutrition, and function of carotenoids. *Annu. Rev. Nutr.* **1986**, *6*, 273–297. [CrossRef]
13. Etievant, P.X. *Wine. Volatile Compounds in Foods and Beverages*; Maarse, H., Ed.; Marcel Dekker: New York, NY, USA, 1991; pp. 483–546.
14. Baumes, R.; Wirth, J.; Bureau, S.; Gunata, Y.; Razungles, A. Biogeneration of C13-norisoprenoid compounds: Experiments supportive for an apo-carotenoid pathway in grapevines. *Anal. Chim. Acta* **2002**, *458*, 3–14. [CrossRef]

15. Young, P.R.; Eyeghe-Bickong, H.A.; du Plessis, K.; Alexandersson, E.; Jacobson, D.A.; Coetzee, Z.; Deloire, A.; Vivier, M.A. Grapevine plasticity in response to an altered microclimate: Sauvignon Blanc modulates specific metabolites in response to increased berry exposure. *Plant Physiol.* **2016**, *170*, 1235–1254. [CrossRef]

16. Ryona, I.; Sacks, G.L. Behavior of glycosylated monoterpenes, c13-norisoprenoids, and benzenoids in Vitis vinifera cv. Riesling during ripening and following hedging. In *Carotenoid Cleavage Products*; American Chemical Society: Washington, DC, USA, 2013; Volume 1134, pp. 109–124. ISBN 0-8412-2778-0.

17. Gunata, Z. Biosynthesis of C13-Norisoprenoids in Vitis vinifera: Evidence of carotenoid cleavage dioxygenase (ccd) and secondary transformation of norisoprenoid compounds. In *Carotenoid Cleavage Products*; American Chemical Society: Washington, DC, USA, 2013; Volume 1134, pp. 97–107. ISBN 0-8412-2778-0.

18. Razungles, A.; Bayonove, C.L.; Cordonnier, R.E.; Sapis, J.C. Grape carotenoids: Changes during the maturation period and localization in mature berries. *Am. J. Enol. Vitic.* **1988**, *39*, 44–48.

19. Crupi, P.; Coletta, A.; Antonacci, D. Analysis of carotenoids in grapes to predict norisoprenoid varietal aroma of wines from Apulia. *J. Agric. Food Chem.* **2010**, *58*, 9647–9656. [CrossRef] [PubMed]

20. Lutz, A.; Winterhalter, P. Bio-oxidative cleavage of carotenoids: Important route to physiological active plant constituents. *Tetrahedron Lett.* **1992**, *33*, 5169–5172. [CrossRef]

21. Oliveira, C.; Silva Ferreira, A.C.; Mendes Pinto, M.; Hogg, T.; Alves, F.; Guedes de Pinho, P. Carotenoid compounds in grapes and their relationship to plant water status. *J. Agric. Food Chem.* **2003**, *51*, 5967–5971. [CrossRef]

22. Gerdes, S.M.; Winterhalter, P.; Ebeler, S.E. Effect of sunlight exposure on norisoprenoid formation in White Riesling grapes. In *Carotenoid-Derived Aroma Compounds*; American Chemical Society: Washington, DC, USA, 2002; pp. 262–272. [CrossRef]

23. Oliveira, C.; Ferreira, A.C.; Costa, P.; Guerra, J.; Guedes de Pinho, P. Effect of some viticultural parameters on the grape carotenoid profile. *J. Agric. Food Chem.* **2004**, *52*, 4178–4184. [CrossRef] [PubMed]

24. Marais, J.; van Wyk, C.J.; Rapp, A. Carotenoid levels in maturing grapes as affected by climatic regions, sunlight and shade. *S. Afr. J. Enol. Vitic.* **1991**, *12*, 64–69. [CrossRef]

25. Razungles, A.J.; Babic, I.; Sapis, J.C.; Bayonove, C.L. Particular behaviour of epoxy xanthophylls during veraison and maturation of grape. *J. Agric. Food Chem.* **1996**, *44*, 3821–3825. [CrossRef]

26. Razungles, A.J.; Baumes, R.L.; Dufour, C.; Sznaper, C.N.; Bayonove, C.L. Effect of sun exposure on carotenoids and C(13)-norisoprenoid glycosides in Syrah berries (Vitis vinifera L.). *Sci. Aliment.* **1998**, *18*, 361–373.

27. Schultz, H.R.; Lohnertz, O.; Bettner, W.; Balo, B.; Linsenmeier, A.; Jahnish, A.; Muller, M.; Gaubatz, B.; Varadi, G. Is grape composition affected by current levels of UV-B radiation? *VITIS* **1998**, *37*, 191–192.

28. Giovanelli, G.; Brenna, O.V. Evolution of some phenolic components, carotenoids and chlorophylls during ripening of three Italian grape varieties. *Eur. Food Res. Technol* **2007**, *225*, 145–150. [CrossRef]

29. Guidoni, S.; Ferrandino, A.; Novello, V. Effects of seasonal and agronomical practices on skin anthocyanin profile of Nebbiolo grapes. *Am. J. Enol. Vitic.* **2008**, *59*, 22–29.

30. Guidoni, S.; Cavalletto, S.; Bartolomei, S.; Mania, E.; Gangemi, L. Microclimatic aspects in vineyards with different vigour and exposure. *Progrès Agricole et Viticole* **2011**, 547–550.

31. Asproudi, A.; Petrozziello, M.; Cavalletto, S.; Guidoni, S. Grape aroma precursors in cv. Nebbiolo as affected by vine microclimate. *Food Chem.* **2016**, *211*, 947–956. [CrossRef] [PubMed]

32. Chorti, E.; Guidoni, S.; Ferrandino, A.; Novello, V. Effect of different cluster sunlight exposure levels on ripening and anthocyanin accumulation in Nebbiolo grapes. *Am. J. Enol. Vitic.* **2010**, *61*, 23–30.

33. Guidoni, S.; Asproudi, A.; Mania, E.; Cavaletto, S.; Gangemi, L.; Matese, A.; Primicerio, J.; Borsa, D. Vineyard vigour, microclimate and secondary metabolites in cv Nebbiolo. *Cienc Tec Vitivinic* **2013**, *28*, 644–648.

34. Crupi, P.; Milella, R.A.; Antonacci, D. Simultaneous HPLC-DAD-MS (ESI+) determination of structural and geometrical isomers of carotenoids in mature grapes. *J. Mass Spectrom.* **2010**, *45*, 971–980. [CrossRef]

35. Crupi, P.; Coletta, A.; Milella, R.A.; Palmisano, G.; Baiano, A.; La Notte, E.; Antonacci, D. Carotenoid and chlorophyll-derived compounds in some wine grapes grown in Apulian region. *J. Food Sci.* **2010**, *75*, S191–S198. [CrossRef]

36. Kamffer, Z.; Bindon, K.A.; Oberholster, A. Optimization of a method for the extraction and quantification of carotenoids and chlorophylls during ripening in grape berries (*Vitis vinifera* cv. Merlot). *J. Agric. Food Chem.* **2010**, *58*, 6578–6586. [CrossRef]

37. Oliveira, C.; Barbosa, A.; Ferreira, A.C.S.; Guerra, J.; Guedes DE Pinho, P. Carotenoid profile in grapes related to aromatic compounds in wines from Douro Region. *J. Food Sci.* **2006**, *71*, S1–S7. [CrossRef]

38. Razungles, A.; Bayonove, C.L.; Cordonnier, R.E.; Baumes, R.L. Etude des caroténoïdes du raisin a maturité. *VITIS* **1987**, *26*, 183–191.
39. Rodriguez-Amaya, D.B.; Kimura, M.; Godoy, H.T.; Amaya-Farfan, J. Updated Brazilian database on food carotenoids: Factors affecting carotenoid composition. *J. Food Compos. Anal.* **2008**, *21*, 445–463. [CrossRef]
40. Dall'Osto, L.; Lico, C.; Alric, J.; Giuliano, G.; Havaux, M.; Bassi, R. Lutein is needed for efficient chlorophyll triplet quenching in the major LHCII antenna complex of higher plants and effective photoprotection in vivo under strong light. *BMC Plant. Biol.* **2006**, *6*, 32. [CrossRef]
41. Bureau, S.M.; Razungles, A.J.; Baumes, R.L.; Bayonove, C.L. Effect of vine or bunch shading on the carotenoid composition in Vitis vinifera L. berries. I. Syrah grapes. *Wein-Wissenschaft* **1998**, *53*, 64–71.
42. Bindon, K.A.; Dry, P.R.; Loveys, B.R. Influence of plant water status on the production of c13-norisoprenoid precursors in *Vitis vinifera* L. Cv. Cabernet sauvignon grape berries. *J. Agric. Food Chem.* **2007**, *55*, 4493–4500. [CrossRef]

© 2020 by the authors. Licensee MDPI, Basel, Switzerland. This article is an open access article distributed under the terms and conditions of the Creative Commons Attribution (CC BY) license (http://creativecommons.org/licenses/by/4.0/).

Article

How Pre-Harvest Inactivated Yeast Treatment May Influence the Norisoprenoid Aroma Potential in Wine Grapes

Pasquale Crupi [1,*], Marika Santamaria [1], Fernando Vallejo [2], Francisco A. Tomás-Barberán [2], Gianvito Masi [1], Angelo Raffaele Caputo [1], Fabrizio Battista [3] and Luigi Tarricone [1]

[1] CREA-VE, Council for Agricultural Research and Economics—Research Centre for Viticulture and Enology, Via Casamassima 148, 70010 Turi (BA), Italy; marika.santamaria@crea.gov.it (M.S.); gianvito.masi@crea.gov.it (G.M.); angeloraffaele.caputo@crea.gov.it (A.R.C.); luigi.tarricone@crea.gov.it (L.T.)

[2] CEBAS-CSIC, Centro de Edafología y Biología Aplicada del Segura, E-30100 Murcia, Spain; fvallejo@cebas.csic.es (F.V.); fatomas@cebas.csic.es (F.A.T.-B.)

[3] LALLEMAND Italia, Via Rossini 14/B, 37060 Castel D'Azzano (VR), Italy; fbattista@lallemand.com

* Correspondence: pasquale.crupi@crea.gov.it; Tel.: +39-080-8915711

Received: 11 April 2020; Accepted: 6 May 2020; Published: 13 May 2020

Abstract: Carotenoids are important secondary metabolites in wine grapes and play a key role as potential precursors of aroma compounds (i.e., C_{13}-norisoprenoids), which have a high sensorial impact in wines. There is scarce information about the influence of pre-harvest inactivated yeast treatment on the norisoprenoid aroma potential of grapes. Thus, this work aimed to study the effect of the foliar application of yeast extracts (YE) to Negro Amaro and Primitivo grapevines on the carotenoid content during grape ripening and the difference between the resulting véraison and maturity (ΔC). The results showed that β-carotene and (all$E$)-lutein were the most abundant carotenoids in all samples, ranging from 60% to 70% of total compounds. Their levels, as well as those of violaxanthin, (9′Z)-neoxanthin, and 5,6-epoxylutein, decreased during ripening. This was especially observed in treated grapes, with ΔC values from 2.6 to 4.2-fold higher than in untreated grapes. Besides this, a principal components analysis (PCA) demonstrated that lutein, β-carotene, and violaxanthin and (9′Z)-neoxanthin derivatives principally characterized Negro Amaro and Primitivo, respectively. Thereby, the YE treatment has proved to be effective in improving the C_{13}-norisoprenoid aroma potentiality of Negro Amaro and Primitivo, which are fundamental cultivars in the context of Italian wine production.

Keywords: HPLC-DAD-MS; 5,6-/5,8-epoxyxanthophylls; elicitors; pheophytins; chlorophylls

---

1. Introduction

The presence of carotenoids in grape berries is well documented [1]. 5,6-epoxyxanthophylls and their 5,8-epoxy isomers were identified together with the most common carotenes and xanthophylls (i.e., β-carotene and lutein), the content of which was established to decrease during grape ripening from véraison to harvest [2–4]. The grape variety and viticulture practices, but also climate conditions and geographic origin, can influence the qualitative and quantitative profile of carotenoids in berries [5–7].

Structurally, carotenoids are C_{40} tetraterpenoids with a long chromophore of conjugated double bonds; thus, they can confer from red to yellow coloration to fruit [8]. However, especially in red/black grapes, this function is ascribed to anthocyanins and polyphenols [9]. Instead, carotenoids are mainly known as precursors of volatile compounds (i.e., C_{13}-norisoprenoids), with very low olfactory perception thresholds. Thus, they have a fundamental role in defining the varietal aroma of grapes and wines [10,11]. C_{13}-norisoprenoids form by the direct enzymatic oxidation of carotenoids, such as β-carotene, lutein, neoxanthin, and violaxanthin, followed by the acid hydrolysis of glycosylated intermediates [12].

There is no standard extraction procedure of carotenoids from foodstuffs; generally, in the case of grapes, aprotic and poor-polarity organic solvents, such as acetone [2] or diethyl ether/hexane 1:1 [13], are employed to extract carotenes and xanthophylls simultaneously. Reversed-phase HPLC coupled with various detection techniques (e.g., DAD and MS) is currently the method of choice for carotenoid analysis [14]. In particular, the C_{30} stationary phase has the highest separation selectivity including structural and geometrical isomers [15].

Currently, the vineyard application of elicitors (i.e., methyl jasmonate or yeast extracts) represents the most original strategy to activate the biosynthesis of secondary metabolites because it may stimulate an innate immune response in grapes [16,17]. Generally, it works in the case of the phenolic and volatile composition of grapes [9,18,19], but recent research has demonstrated that treatment with methyl jasmonate and yeast extracts was also determinant for increasing lutein and β-carotene concentrations in Tempranillo, Graciano, and Garnacha [20]. Additionally, the enzymes involved in carotenoid biosynthesis and degradation are acknowledged to be stress-dependent [21,22].

A previous study conducted by our research group on carotenoid degradation during the grape ripening of four varieties (Chardonnay, Merlot, Primitivo, and Negro Amaro) in the Apulia region proposed a parameter (ΔC) to distinguish grapes with higher C_{13}-norisoprenoid aroma potentiality. ΔC (μg/kg), calculated as the difference between the total carotenoid content at véraison and harvest, was lowest in the autochthonous Primitivo and Negro Amaro [7]. Considering the supposed effect of yeast extracts on carotenoids, their application to vineyards could be hypothesized to enhance ΔC in grapes, although, to the best of our knowledge, no other reports exist in the literature about this property.

Therefore, this work aimed to assess the effectiveness of pre-harvest treatment by inactivated yeast extracts (YE) in improving the C_{13}-norisoprenoid aroma potentiality of Apulian Negro Amaro and Primitivo varieties, which are fundamental in the context of Italian wine production.

2. Materials and Methods

2.1. Plant Materials

The experiment was conducted in 2019 on a 10 year-old commercial vineyard of Negro Amaro *cv.* located in the Salice Salentino wine grape area (Masseria Filippi, Apulia region, Southern Italy) and a 6 year-old commercial vineyard of Primitivo *cv.* located in the Gioia del Colle D.OC. (Denomination of Controlled Origin) area (Azienda Benagiano, Apulia region, Southern Italy). Both Negro Amaro and Primitivo vines, grafted onto *Vitis berlandieri* × *Vitis rupestris* 1103 Paulsen rootstock, were planted in north–south oriented rows; they were spaced 2.15 m between rows and 0.90 m on the row, trained to a Vertical Shoot Positioned (VSP) system and spur-pruned with 14 buds per vine.

The experimental design for the current study was a randomized complete block with four replicates, with three rows of each treatment (80 vines per row) separated by a buffer row. Treatment consisted in two canopy spraying times, the first at the beginning of veráison (5%) and the second 12 days later, with Lalvigne® MATURE diluted in water without adjuvant at 1 kg/ha dose (YE) in comparison to untreated rows used as control (C). Lalvigne® MATURE is a formulation of 100% natural inactivated wine yeast (*Saccharomyces cerevisiae*) derivatives, non-pathogenic, non-hazardous, food grade and non-GMO, which are specifically designed to be used with the patented foliar application technology WO/2014/024039 (Lallemand Inc., Montreal, QC, Canada).

A random sample of three bunches for each repetition was manually picked between the third and seventh node from random vines at periodic intervals from August until September. Berries (~30 g) were then chosen at random from these bunches and stored in the dark at −80 °C until analysis.

2.2. Chemicals

HPLC-grade hexane and acetone were purchased from J.T. Baker (Deventer, Holland). LCMS grade water, methanol and tert-butyl-methyl-ether were purchased from Chromasolv (Exacta+Optech Labcenter S.p.A., Modena, Italy). Sodium hydroxide (NaOH) 0.1 N and bromothymol blue were purchased from Sigma Aldrich (Merck Life Science S.r.l., Milano, Italy). β-apo-8′-carotenal and 3-tert-butyl-4-hydroxyanisole (BHA) were purchased from Fluka (Exacta+Optech Labcenter S.p.A., Modena, Italy). β-carotene and magnesium carbonate basic were purchased from Sigma-Aldrich (Merck Life Science S.r.l., Milano, Italy); (all*E*)-lutein and zeaxanthin were obtained from Extrasynthese (Genay, France), whereas (9′*Z*)-neoxanthin, violaxanthin, 5,6-epoxide-lutein, and (9*Z*)-β-carotene were obtained from CaroteNature (Münsingen, Switzerland) and used as HPLC reference standards.

2.3. Maturation Indexes of Grapes

Total soluble solids (TSS), titratable acidity (TA), and pH were determined according to protocols established by the OIV, 1990 [23]. Berries were crushed to determine the TSS (expressed as g/L) of berry juice using a portable refractometer (ATAGO PR32). TA (as g/L of tartaric acid equivalents) was also determined for the juice by titrating with 0.1 N sodium hydroxide to the bromothymol blue end point. Finally, juice pH was measured by the pH meter CRISON BASIC 20.

2.4. Extraction of Carotenoids from Grapes

The carotenoid extraction procedure was adapted from the method of Crupi et al. 2010 [4]. Approximately 30 g of freeze berries containing 25 μL of BHA (12.66 mg/mL in EtOH) were crushed in an IKA A11 basic homogenizer (IKA, WERKE GMBH & CO.KG, Staufen, Germany) for 5 min in the presence of magnesium carbonate basic (2 to 4 g for mature and green berries, respectively). The homogenate was diluted with 40 mL of distillated water and spiked with 100 μL of internal standard (150 μg/mL of β-apo-8′-carotenal). Extraction was done with 40 mL of hexane/diethyl ether (1:1, *v/v*) by agitating the mixture for 30 min. The resulting upper layer was separated by centrifugation at 4000 rpm and 4 °C for 1 min (EPPENDORF 5810R, Eppendorf AG, Hamburg, Germany). The extraction procedure was repeated twice for the lower phase using 20 mL of hexane/diethyl ether (1:1, *v/v*). The pooled extract was evaporated to dryness using a rotovapor Buchi-R-205. The residue was dissolved in 2 mL of acetone/hexane (1:1, *v/v*), filtered through 0.20 μm syringe PTFE filters and stored at −20 °C until the carotenoid analysis by HPLC-DAD-MS.

2.5. HPLC-DAD-MS Analyses of Carotenoids

HPLC-DAD-MS analyses were carried out with an Agilent model 1100 equipped with quaternary pump solvent delivery, a thermostated column compartment, diode array detector, and XCT-trap mass detector (Agilent Technologies, Palo Alto, CA, USA). A positive electrospray mode was used for the ionization of molecules with acquisition of mass spectra between m/z 100 and 1200, capillary voltage at −4000 V, and skimmer voltage at 30 V. The reversed stationary phase employed was a YMC pack C_{30} (YMC Inc., Wilmington, NC, USA) 5 μm (250 × 3 mm i.d.) with a precolumn C_{30} 5 μm (20 × 3 mm i.d.). The following gradient system was used with H_2O (solvent A), methanol (solvent B), and tert-butyl methyl ether (solvent C): 0 min, %A–%B–%C, 40–60–0; 5 min, %A–%B–%C, 20–80–0; 10 min, %A–%B–%C, 4–81–15; 60 min, %A–%B–%C, 4–11–85; 65 min, %A–%B–%C, 4–11–85; 70 min, %A–%B–%C, 40–60–0. Stop time at 70 min with a re-equilibration time of 20 min corresponding to ca 3.0 column volumes (Vc = 1.3 mL). The column was kept at 20 °C, the flow was maintained at 0.2 mL/min, and the sample injection volume was 3 μL. Diode array detection was between 250 and 700 nm and absorbance was recorded at 447 nm. Positive electrospray mode was used for the ionization of molecules with a capillary voltage at −4000 V and skimmer voltage at 40 V. The nebulizer pressure was 15 psi and the nitrogen flow rate was 5 L/min. The temperature of drying gas was 350 °C. In the full scan mode, the monitored mass range was from m/z 100 to 1200. MS^2 was performed by using helium as the collision gas at a pressure of 4.6×10^{-6} mbar. Collision induced dissociation (CID) spectra were obtained with an isolation width of 4.0 *m/z* for precursor ions and a fragmentation amplitude of 0.6 V for epoxyxanthophylls, and 1.0 V for the other carotenoids.

Compound identification was achieved by combining different information: the positions of absorption maxima (λ_{max}), the degree of vibration fine structure (% III/II), the ratio of the absorbance of the *cis* peak to the absorbance of the second absorption band in the visible region, known as the Q ratio or D_B/D_{II} [8,24], and the capacity factor values k' and mass spectra were compared with those from pure standards and interpreted with the help of structural models already hypothesized in the literature (such as (9Z) or (9'Z)-lutein and the other lutein-like structures reported in Table 1, which were only tentatively identified) [4].

Table 1. HPLC-DAD-MS (ESI$^+$) characteristics of carotenoids in grapes.

Peak	Compound	k'	λ_{max} (nm)	% (III/II)[a]	D_B/D_{II}[b]	[M+H]$^+$(m/z)	[M]$^+$(m/z)	MS^2 Product Ions m/z
1	violaxanthin like structure	2.78	418;440;470	84		601.5	<u>600.1</u>	583.5, 565.5, 509.5, 491.5, 221.1
2	violaxanthin	2.92	416; 440; 468	86		601.5	<u>600.1</u>	583.5, 565.5, 509.5, 491.5, 221.1
3	(8'R)-neochrome	3.01	400; 422; 450	88		601.5	<u>600.1</u>	583.2, 565.3, 509.5, 221.1
4	(9'Z)-neoxanthin	3.06	414; 436; 464	86		601.5	<u>600.1</u>	583.2, 565.3, 509.5, 221.1
5	(8'S)-neochrome	3.12	400; 422; 450	88		601.5	<u>600.1</u>	583.2, 565.3, 509.5, 221.1
6	5,6-epoxylutein	3.19	416; 440; 468	90		585.4	<u>584.2</u>	567.1, 493.1, 221.1
7	luteoxanthin	3.33	399; 422; 448	94		601.5	<u>600.1</u>	583.2, 221.1
8	lutein like structure	3.42	(425); 446; 474			568.9	567.9	550.9, 532.9, <u>476.4</u>, 429.4
9	Z lutein like structure	3.47	328; (412); 436; 464			568.9	567.9	550.9, 532.9, <u>476.4</u>, 429.4
10	(8'S)-auroxanthin	3.64	380; 402; 426	98		601.5	<u>600.1</u>	583.5, 565.5, 509.5, 491.5, 221.1
11	chlorophyll b	3.70	258; 314; 342; 466; 600; 650					
12	(*allE*)-lutein	3.75	(422); 446; 472	40		568.9	567.9	550.9, 532.9, <u>476.4</u>, 429.4
13	zeaxanthin	4.00	(425); 452; 476	22		568.9	567.9	550.9, 532.9, <u>476.4</u>, 429.4
14	(9Z) or (9'Z)-lutein	4.16	330; (422); 440; 468	50	0.075	568.9	567.9	550.9, 532.9, <u>476.4</u>, 429.4
15	chlorophyll a	4.31	336; (385); (417); 432; 618; 665					
IS	β-apo-8'-carotenal	4.50	460					
16	pheophytin b	5.46	(417); 436; 527; 600; 654				885	
17	pheophytin a	5.59	410; 506; 536; 666				871	
18	β-carotene	5.98	(430); 452; 478	25		536.9	535.9	<u>444.2</u>, 430.3, 399.3
19	(9Z)- β-carotene	6.25	342; (424); 446; 474	17	0.03	536.9	535.9	<u>444.2</u>, 430.3, 399.3

[a] %III/II is the ratio of the height of the longest-wavelength absorption peak, designated III, and that of the middle absorption peak, designated as II, taking the minimum between the two peaks as a baseline [8]; [b] The Q ratio is the quotient between the cis peak band and band II (normally λ_{max}.) [24]. The fragments corresponding to the base peak in MS^2 spectra were underlined.

To evaluate linearity, calibration curves with seven concentration points for each compound were prepared separately. Calibration was performed by a linear regression of peak–area ratios of the carotenoids to the internal standard (β-apo-8′-carotenal) versus the respective standard concentration. The precision of the method was determined by calculating the intraday and interday repeatability, expressed as standard deviation (SD) (σ) and relative standard deviation (RSD) in terms of retention times and peak width ($W_{1/2}$), by the injection of four different replicates of extracts on the same day or over three consecutive days. The detection limit (LOD) and quantification limit (LOQ) were calculated on the basis of the calibration curve of low concentrations of target compounds and defined as LOD = 3*(Syx/b) and LOQ = 10*(Syx/b), where b is the slope and Syx is the standard error of the calibration curve (Supplementary Materials, Table S1).

Depending on the type of chemical structure, carotenoids were quantified as μg violaxanthin (in the case of violaxanthin and a violaxanthin-like structure), (9′Z)-neoxanthin (in the case of (9′Z)-neoxanthin, (8′R)-neochrome, (8S)-neochrome, and auroxanthin), 5,6-epoxylutein (in the case of 5,6-epoxylutein and luteoxanthin), (allE)-lutein (in the case of (allE)-lutein, (9Z)-lutein, and lutein-like structures), zeaxanthin (in the case of zeaxanthin), and β-carotene (in the case of β-carotene and (9Z)-β-carotene) equivalents per kg of berries.

2.6. Statistical Analysis

Data were analyzed by the STATISTICA 8.0 (StatSoft Inc., Tulxa, OK, USA, 2012) software package. Specifically, after testing their normal distribution by Shapiro–Wilk's W test together with their homoscedasticity by means of the Levene test, a two-way multivariate analysis of variance (MANOVA) was performed for the HPLC-DAD-quantified carotenoids in order to evaluate the effects of the factors "treatment" and "ripening" on each cultivar. The Tukey HSD post hoc test was used to separate the means ($p < 0.05$) when the interaction between the factors was significant.

Finally, a principal components analysis (PCA) was performed on variables corresponding to the difference between the values of the identified carotenoids at véraison and maturity to cluster samples per treatment. The PCA was done on the correlation matrix in order to treat all variables on an equal footing, in consecutive steps (as reported by Martì et al. 2004 [25]) and starting from those variables which were shown to be significant through the MANOVA.

3. Results and Discussion

3.1. Carotenoid Composition of the Wine Grape Varieties

Nineteen compounds including carotenoids (15), chlorophylls (2), and pheophytins (2) were detected by HPLC-DAD-MS in the wine grape extracts (Figure 1).

It is worth noting that consistency in carotenoid attribution is very complex due to their structural diversity. In particular, the double bond-conjugated system is susceptible to heat, light, oxygen, and acids, giving rise to unwanted *cis–trans* isomerization. Using antioxidants (i.e., BHA), an ice bath, and minimizing the light exposition of the samples during the extraction steps are mandatory to prevent the formation of *cis*-isomers as artifact compounds [26]. Therefore, alongside the main trans carotenoids identified, β-carotene and (allE)-lutein, accounting for 60% and 70% of total carotenoids in mature Primitivo and Negro Amaro grapes [7], we were also confident that the *cis*-isomers (assigned as 9Z-lutein and 9Z-β-carotene) were present in the grape tissue (Table 1).

Analogously, an acidic medium could cause the isomerization of 5,6-epoxyxanthophylls (Figure 2). Indeed, treating an acetone/hexane solution of violaxanthin and (9′Z)-neoxanthin (3 μg/mL) with 500 μL of tartaric acid (15 g/L) at a concentration similar to that of green berries, hypsochromic shifts of their absorption maxima (λmax) of 20 nm (corresponding to the generation of luteoxanthin and neochrome) and 40 nm (corresponding to the generation of auroxanthin) were observed within 30 min (Figure 3). Thanks to the addition of $(MgCO_3)4Mg(OH)_2$ $5H_2O$, which neutralized the extraction solution, we could exclude that the recognized 5,8-epoxyxanthophylls, (8′R)-neochrome (peak 3),

(8′S)-neochrome (peak 5), luteoxanthin (peak 7) and auroxanthin (peak 10), were derived from the isomerization of violaxanthin (peak 2) and (9′Z)-neoxanthin (peak 4), respectively, during the extraction (Table 1).

Furthermore, the absence of flavoxanthin and chrysantemaxanthin, originating from 5,6-epoxylutein (Figure 2), also confirmed this speculation. Although the lack of these two xanthophylls appeared to disagree with our past report [4], it is worth noting that the carotenoid composition is often influenced by climate and viticulture practices, such as trellis systems [27].

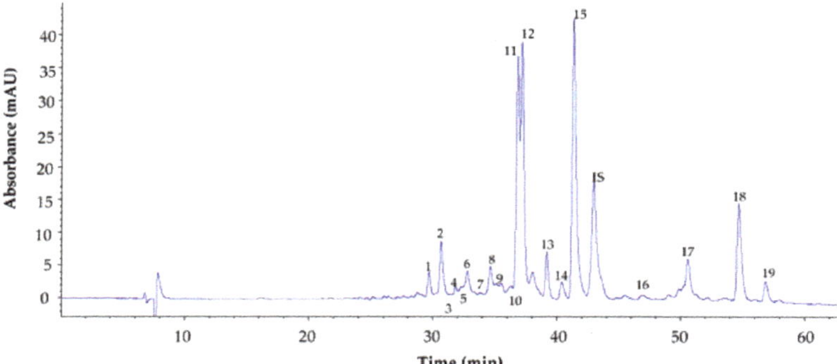

Figure 1. HPLC-DAD profile of carotenoids in mature grapes (Primitivo *cv*). Peaks: (1) violaxanthin-like structure; (2) violaxanthin; (3) (8′R)-neochrome; (4) (9′Z)-neoxanthin; (5) (8′S)-neochrome; (6) 5,6-epoxylutein; (7) luteoxanthin; (8) lutein-like structure; (9) *Z* lutein-like structure; (10) (8′S)-auroxanthin; (11) chlorophyll b; (12) (all*E*)-lutein; (13) zeaxanthin; (14) (9*Z*) or (9′*Z*)-lutein; (15) chlorophyll a; (16) pheophytin b; (17) pheophytin a; (18) β-carotene; (19) (9*Z*)-β-carotene. IS β-apo-8′-carotenal.

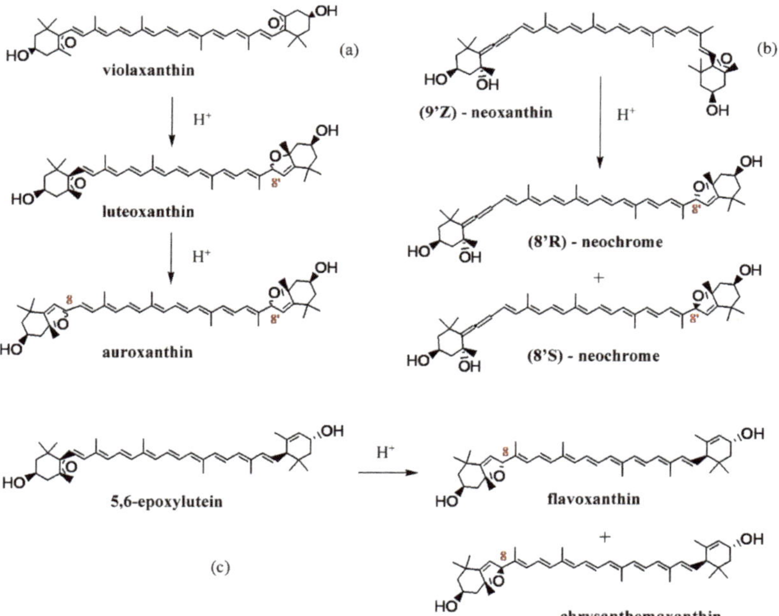

Figure 2. Acid-catalyzed isomerization of (**a**) violaxanthin, (**b**) (9′Z)-neoxanthin, and (**c**) 5,6-epoxylutein.

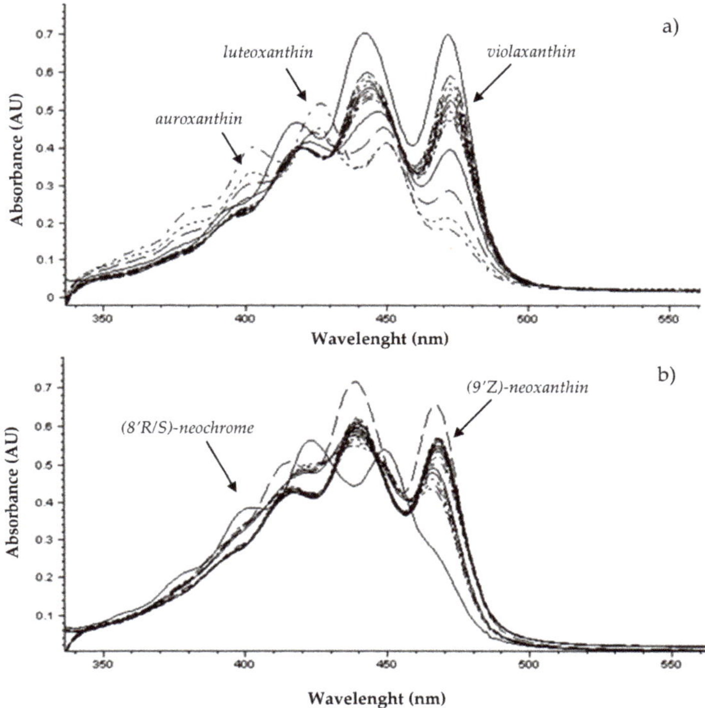

Figure 3. UV-Vis spectra of (**a**) violaxanthin and (**b**) (9′Z)-neoxanthin (3 µg/mL) treated with tartaric acid (15 g/L).

3.2. Effect of Pre-Harvest YE Application on Carotenoid Content

According to the choice of the main figures of merit reported elsewhere [28], the adopted HPLC-DAD method was validated for the target analysis of carotenoids in terms of linearity, LOD, LOQ, and precision (Supplementary Materials, Table S1). The calibration curves were obtained by the internal standard method to give determination coefficients (R^2) which were higher than 0.99; moreover, the detection and quantification limits were found to be under 2 and 5 µg/kg, respectively. As regards precision tests, intra-assay and inter-assay coefficients of variation were calculated, with retention times ranging between 0.11% and 0.13%, and 0.37% and 0.50%, respectively, while corresponding measures for $W_{1/2}$ were between 2.9% and 15%, and 2.8% and 12%, respectively. These results confirmed the accuracy and reproducibility of data obtained by the analytical method for carotenoid analysis.

Tables 2 and 3 listed the changes in carotene and xanthophyll concentrations (expressed in µg/kg of berries) during the grape ripening of the two varieties, Negro Amaro and Primitivo, in control and YE-treated samples. As expected, total carotenoids significantly decreased from véraison to maturity, even more consistently in Negro Amaro (F = 34.06, $p < 0.001$), due to the higher accumulation of the compounds in the first sampling date, respective to Primitivo (F = 19.54, $p < 0.001$) (Figure 4). Regarding the effect of the elicitor (YE), the treatment factor (T) as well as its interaction with ripening (TxR) resulted in a more pronounced effect in the variation of total carotenoid content in Primitivo (T: F = 29.11, $p < 0.001$; TxR: F = 7.06, $p < 0.001$) than Negro Amaro (T: F = 5.55, $p = 0.0047$; TxR: F = 6.05, $p = 0.0021$). This difference was mostly influenced by the anomalous behavior of the P0813 control sample; the lowest carotenoid value could be ascribed to an experimental drawback that occurred during the sampling and/or extraction phase, as suggested by the extraction yields in the four replicates, which were low (Supplementary Materials, Figure S1).

Table 2. Changes in carotenoid contents (expressed as µg/kg of berries) in Negro Amaro grapes from véraison to maturity.

Sampling Date	08/13/19*		08/26/19		09/02/19		09/17/19		ΔC^a	
	C	YE	C	YE	C	YE	C	YE	C	YE
TSS[b]	153.0[c] ± 1.7[d]	142.3 ± 1.2	165 ± 2	163 ± 2	171.4 ± 1.4	189 ± 1.0	212 ± 3	220 ± 3		
pH	3.11 ± 0.01	3.17 ± 0.01	3.10 ± 0.02	3.26 ± 0.01	3.31 ± 0.06	3.38 ± 0.04	3.36 ± 0.04	3.43 ± 0.06		
TA[e]	8.47 ± 0.15	7.78 ± 0.16	6.25 ± 0.18	5.82 ± 0.10	5.63 ± 0.07	6.6 ± 0.4	5.8 ± 0.3	5.4 ± 0.4		
TSS/TA[f]	18.1 ± 0.4	18.3 ± 0.5	26.4 ± 1.1	27.9 ± 0.4	30.4 ± 0.2	29 ± 3	37 ± 2	40 ± 3		
Compounds										
violaxanthin like structure[g]	30 ± 5c[h]	64 ± 15a	49 ± 5ab	34 ± 5bc	39 ± 6bc	42 ± 7bc	34 ± 5bc	29 ± 4c		
violaxanthin[g]	96 ± 10b	126 ± 14a	80 ± 12bc	67 ± 10c	91 ± 14bc	97 ± 15b	79 ± 12bc	68 ± 9c		
(8′R)-neochrome[i]	8.2 ± 0.6a	9.5 ± 0.3a	8.5 ± 0.8a	4.8 ± 1.0b	3.2 ± 0.6c	4.2 ± 0.4bc	2.8 ± 0.5c	2.95 ± 0.17c		
(9′Z)-neoxanthin[i]	38 ± 10ab	56 ± 10a	58 ± 12a	50 ± 8ab	47 ± 13ab	45 ± 6ab	39 ± 7ab	31 ± 3b		
(8′S)-neochrome[i]	6.0 ± 0.9b	9 ± 2a	7.3 ± 1.3ab	6.2 ± 1.1bc	4.6 ± 0.7c	5.5 ± 0.8bc	4.0 ± 0.6c	3.8 ± 0.5c		
5,6-epoxy-lutein[l]	68 ± 5a	82 ± 8a	53 ± 9b	39 ± 6bc	44 ± 7bc	43 ± 6bc	38 ± 6bc	30 ± 3c		
luteoxanthin[g]	8.8 ± 1.2b	14 ± 2a	8.2 ± 1.0bc	7.6 ± 1.5bc	9.0 ± 1.0b	5.8 ± 0.8c	7.8 ± 0.9bc	tr		
lutein like structure[m]	44 ± 7b	62 ± 5a	34 ± 5bc	32 ± 5bcd	29 ± 6cd	29 ± 7cd	25 ± 6cd	20 ± 4d		
Z lutein like structure[m]	9.8 ± 0.8b	14.9 ± 1.8a	tr	tr	5.9 ± 1.2c	4.1 ± 1.8cd	5.1 ± 1.1cd	2.8 ± 1.2d		
(8′S)-auroxanthin[g]	54 ± 7a	68 ± 15a	29 ± 5b	10 ± 3c	19 ± 8bc	18 ± 3bc	15 ± 5bc	12 ± 2bc		
(all-E)-lutein[m]	600 ± 90b	780 ± 60a	570 ± 90bc	490 ± 90bcd	410 ± 60cd	390 ± 70d	360 ± 60de	200 ± 20e		
zeaxanthin[n]	88 ± 4	101 ± 5	65 ± 9	65 ± 15	30 ± 5	28 ± 4	26 ± 5	20 ± 2		
(9Z)-lutein[m]	17 ± 5c	33 ± 5a	30 ± 6ab	20 ± 5bc	10 ± 3d	11 ± 3cd	9 ± 2d	7 ± 2d		
β-carotene[o]	390 ± 60bc	420 ± 30abc	390 ± 60bc	310 ± 60c	550 ± 90a	390 ± 60bc	470 ± 80ab	270 ± 40c		
(9Z)-β-carotene[o]	130 ± 50	110 ± 20	180 ± 50	150 ± 40	230 ± 20	167 ± 19	200 ± 20	116 ± 8		
Total[p]	1600 ± 170ab	1950 ± 90a	1600 ± 200ab	1280 ± 180b	1500 ± 200b	1280 ± 180b	1320 ± 170b	820 ± 30c	270 ± 300	1130 ± 120

[a] Difference of total carotenoid concentrations between véraison and maturity; [b] Total soluble solids are expressed in g/L; [c] Means of three replicates; [d] Standard deviation at $p \leq 0.05$; [e] Total acidity expressed in g/L as tartaric acid; [f] Maturation index; [g] Expressed as violaxanthin equivalent; [h] Different letters in the same line are significantly different at the 5% level (Tukey's Honestly Significant Difference test); [i] Expressed as (9′Z)-neoxanthin equivalent; [l] Expressed as 5,6-epoxy-lutein equivalent; [m] Expressed as (allE)-lutein equivalent; [n] Expressed as zeaxanthin equivalent; [o] Expressed as β-carotene equivalent; [p] Sum of identified carotenoids. * veraison: phenologic phase 10% of berries softening and/or coloring. tr: Present at a concentration lower than the limit of quantification (LOQ).

Table 3. Changes in carotenoid contents (expressed as μg/kg of berries) in Primitivo grapes from véraison to maturity.

Sampling Date		08/13/19		08/26/19*		09/04/19		09/19/19		ΔC^a	
		C	YE	C	YE	C	YE	C	YE	C	YE
TSS[b]		$175^c \pm 1.0^d$	170 ± 2	187 ± 1.5	200 ± 2	226 ± 4	227 ± 14	236 ± 5	239 ± 15		
pH		2.98 ± 0.02	2.97 ± 0.03	3.19 ± 0.02	3.22 ± 0.02	3.40 ± 0.01	3.35 ± 0.08	3.45 ± 0.01	3.40 ± 0.07		
TA[e]		11.2 ± 0.4	13.1 ± 0.6	6.8 ± 0.3	7.03 ± 0.15	6.2 ± 0.4	6.7 ± 0.3	5.7 ± 0.4	6.1 ± 0.4		
TSS/TA[f]		15.7 ± 0.5	12.8 ± 0.4	27.8 ± 1.3	28. 8 ± 0.7	36.3 ± 1.9	34 ± 3	41 ± 2	39 ± 3		
Compounds											
violaxanthin like structure[g]		$5 \pm 2d^h$	21 ± 6bc	39 ± 9a	25 ± 7abc	35 ± 7ab	22 ± 6bc	30 ± 6abc	17 ± 5cd		
violaxanthin[g]		31 ± 19d	144 ± 6ab	160 ± 30a	129 ± 9ab	154 ± 18ab	115 ± 17bc	132 ± 15ab	87 ± 12c		
(8′R)-neochrome[i]		tr	tr	tr	3.2 ± 0.5	tr	tr	tr	tr		
(9′Z)-neoxanthin[i]		4.7 ± 1.4d	50 ± 15ab	56 ± 15ab	72 ± 17a	45 ± 11b	19 ± 5cd	39 ± 9bc	15 ± 4cd		
(8′S)-neochrome[i]		10 ± 3a	3.4 ± 1.6b	2.3 ± 0.6b	5.3 ± 1.3b	3.5 ± 1.2b	3.3 ± 0.7b	3.0 ± 1.0b	2.5 ± 0.5b		
5,6-epoxy-lutein[l]		33 ± 13d	79 ± 9abc	100 ± 20ab	94 ± 14ab	104 ± 10a	74 ± 7bc	89 ± 8ab	56 ± 5cd		
luteoxanthin[g]		6 ± 2	tr	tr	tr	tr	tr	tr	tr		
lutein like structure [m]		18 ± 12d	86 ± 12a	74 ± 13ab	69 ± 8ab	61 ± 10b	36 ± 4cd	52 ± 9bc	28 ± 3d		
Z lutein like structure [m]		tr	19 ± 2ab	25 ± 2a	18 ± 4ab	24 ± 5a	18 ± 4ab	21 ± 4ab	14 ± 3b		
(8′S)-auroxanthin[g]		16 ± 3a	11 ± 2abc	10 ± 5abc	15 ± 5ab	12.3 ± 1.3abc	8.1 ± 1.5bc	10.5 ± 1.1abc	6.1 ± 1.1c		
(all-E)-lutein[m]		250 ± 110d	610 ± 90a	610 ± 80a	610 ± 70a	480 ± 40ab	329 ± 19cd	410 ± 30bc	249 ± 14d		
zeaxanthin[n]		30 ± 20c	110 ± 20a	68 ± 16b	64 ± 6b	39 ± 12bc	18 ± 3c	33 ± 10c	13 ± 2c		
(9Z)-lutein[m]		9 ± 7bc	27 ± 4a	26 ± 6a	20 ± 3ab	15 ± 6bc	7 ± 5c	13 ± 5bc	5 ± 4c		
β-carotene[o]		170 ± 110c	320 ± 40b	380 ± 80ab	460 ± 70ab	497 ± 19a	480 ± 40ab	426 ± 17ab	370 ± 30ab		
(9Z)-β-carotene[o]		69 ± 4	51 ± 6	57 ± 10	52 ± 11	48 ± 6	51 ± 11	41 ± 5	39 ± 9		
Total[p]		700 ± 300d	1500 ± 180ab	1600 ± 200a	1640 ± 120a	1520 ± 80ab	1190 ± 80bc	1310 ± 70ab	900 ± 60cd	290 ± 300	740 ± 180

[a] Difference of total carotenoid concentrations between véraison and maturity; [b] Total soluble solids are expressed in g/L; [c] Means of three replicates; [d] Standard deviation at $p \leq 0.05$; [e] Total acidity expressed in g/L as tartaric acid; [f] Maturation index; [g] Expressed as violaxanthin equivalent; [h] Different letters in the same line are significantly different at the 5% level (Tukey's Honestly Significant Difference test); [i] Expressed as (9′Z)-neoxanthin equivalent; [l] Expressed as 5,6-epoxy-lutein equivalent; [m] Expressed as (allE)-lutein equivalent; [n] Expressed as zeaxanthin equivalent; [o] Expressed as β-carotene equivalent; [p] Sum of identified carotenoids. * véraison: phenologic phase 10% of berries softening and/or coloring; because of the experimental drawback mentioned in the text, it was considered at the second sampling point. tr: present at concentration lower than LOQ.

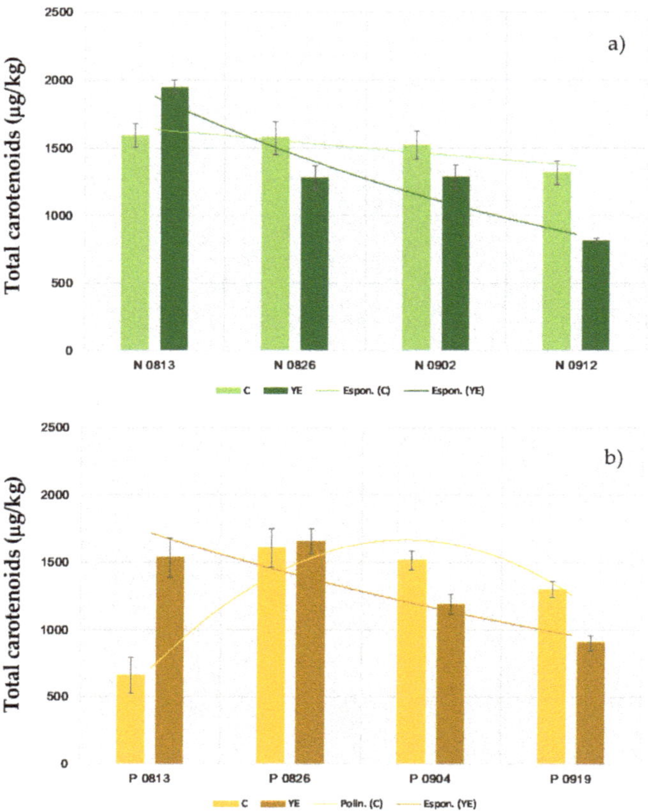

Figure 4. Total carotenoid variation in yeast extract-treated (YE) and control (C) Negro Amaro (**a**) and Primitivo (**b**) wine grapes during ripening.

β-carotene and (all*E*)-lutein were the main compounds in terms of concentration in all analyzed samples. The xanthophyll level was reduced more drastically than the carotene level, reaching values down to 60%–70% in mature grapes compared to the setting (Tables 2 and 3). Moreover, our data clearly showed that the at-harvest β-carotene content was greater than lutein in both varieties, which was in accordance with our previous report [7]. Geographic origin is recognized to affect the relative (all*E*)-lutein and β-carotene contents [2]. In some regions (northern Spain and southern Italy), β-carotene prevailed in grapes at maturity [7,20] in contrast to other cultural situations (northern Italy and southern France) in which the xanthophyll level was higher than the carotene level [1,3].

The interaction between treatment and ripening was very significant for lutein variation ($p < 0.001$) in both varieties (Figure 5a). YE-treated grapes showed an enhanced content of lutein only at véraison; then, they had lower levels in the following samples until reaching 200 µg/kg and 249 µg/kg of berries in Negro Amaro and Primitivo, respectively (Tables 2 and 3). An oscillating trend was observed in the case of β-carotene (Figure 5b). YE treatment conditioned the level of variation of this carotenoid less homogenously in Negro Amaro (TxR: F = 5.04, $p = 0.008$) and, especially, in Primitivo (TxR: F = 4.43, $p = 0.013$). However, a common reduction of the compound between YE (270 µg/kg and 370 µg/kg) and control samples (470 µg/kg and 426 µg/kg) was recorded in both cultivars (Tables 2 and 3). These findings were in disagreement with those of Gutierrez-Gamboa et al., 2018 [20], who reported that YE foliar application triggered β-carotene and lutein accumulation in mature grapes; however, they studied its effect on different varieties (Tempranillo, Graciano, and Garnacha).

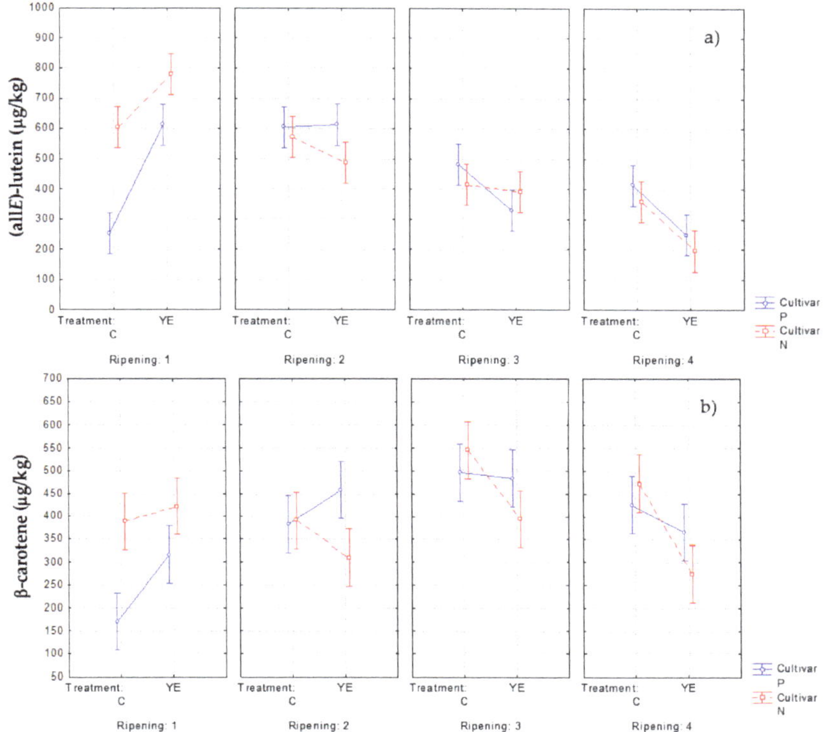

Figure 5. (all*E*)-lutein (**a**) and β-carotene (**b**) contents in yeast extract-treated (YE) and control (C) Negro Amaro and Primitivo wine grapes during ripening.

Finally, concerning the identified 5,6-epoxyxanthophylls, the data confirmed the difficulty of defining the border between the end of storage and the beginning of degradation, mainly in the case of 5,6-epoxylutein (Figure 6c) [7]. Nevertheless, the elicitor treatment caused a significant reduction in the concentrations of violaxanthin, (9′*Z*)-neoxanthin, and 5,6-epoxylutein, as particularly evident in the two last sampling dates of Primitivo grapes (Figure 6; Tables 2 and 3).

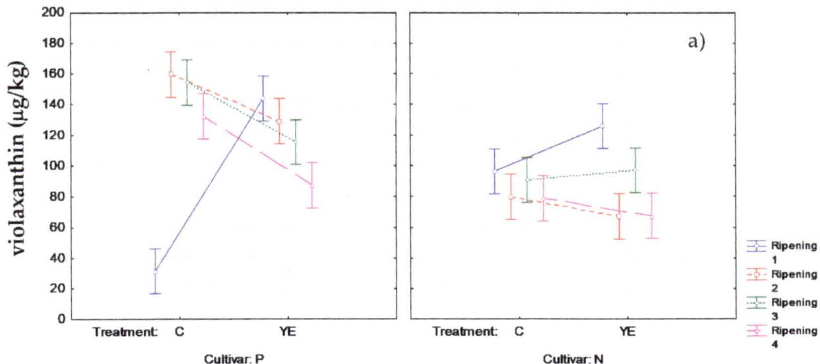

Figure 6. *Cont.*

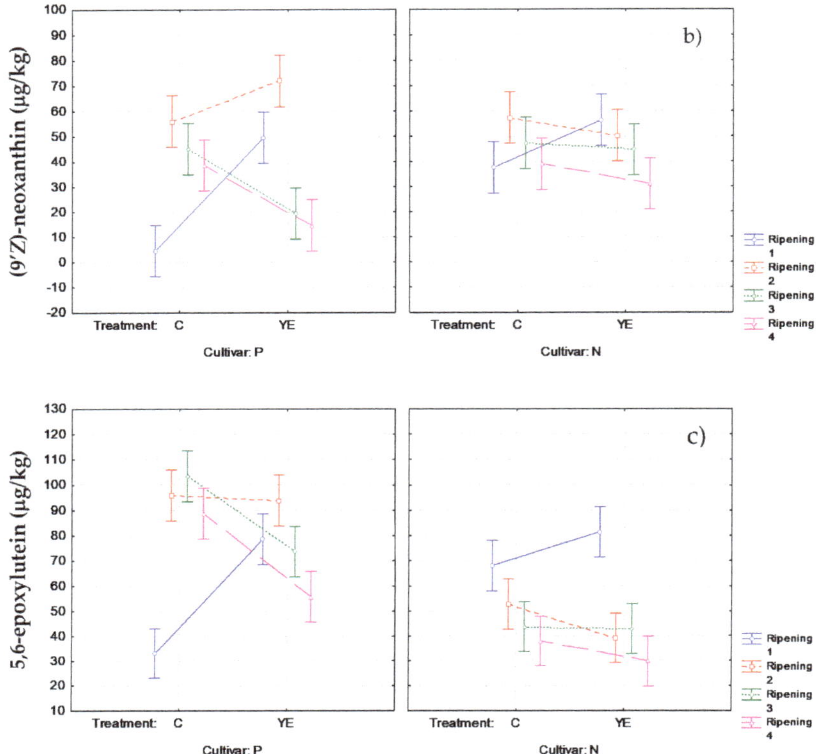

Figure 6. Violaxanthin (**a**), (9′Z)-neoxanthin (**b**), and 5,6-epoxylutein (**c**) contents in yeast extract-treated (YE) and control (C) Negro Amaro and Primitivo wine grapes during ripening.

3.3. Influence of YE Treatment on the Norisoprenoid Aroma Potential (ΔC)

As accepted, C_{13}-norisoprenoid varietal aroma in wine grapes is strictly related to carotenoid degradation during ripening [5,29]. Therefore, determining the difference of total carotenoids between véraison and harvest points (ΔC) together with their decreasing kinetics can be a useful tool for defining the aroma potentiality of grapes for wine making [7]. On the other side, elicitor treatment by foliar application in vineyards has proved to influence the development of aroma compounds in wines [19]; in particular, grapevines treated with benzothiadiazole have been shown to increase the concentrations of some C_{13}-norisoprenoids, such as β-damascenone and β -ionone, while the use of methyl jasmonate did not alter their content [30].

In our work, a similar change in norisoprenoid aroma potential was shown by control samples of Negro Amaro and Primitivo, with ΔC values of 270 µg/kg and 290 µg/kg, respectively (Tables 2 and 3). This finding was in agreement with data from a 3 year-long study conducted on the same cultivars in Apulia [7]. Moreover, the YE treatment significantly affected the aroma potential of the grapes, increasing the ΔCs 2.6 and 4.2-fold in Primitivo (740 µg/kg) and Negro Amaro (1130 µg/kg), respectively (Tables 2 and 3). In the literature, it is well documented that the use of elicitors (i.e., methyl jasmonate, chitosan, yeast extracts) stimulates plant defense mechanisms and consequently activates the enzymes responsible for the biosynthesis of secondary metabolites, such as phenolic compounds [16,31]. YE caused an accumulation of anthocyanins, stilbenes, and flavonoids in wine grapes without varying their phenological maturation [18,32,33]. Since there was no significant difference in maturation indexes between YE and C samples (Tables 2 and 3), the effect of treatment on ΔC variation could be attributed to the activation of enzymes catalyzing either the biosynthesis of carotenoids (as proved by the highest values being in YE grapes at véraison) or their degradation (as confirmed by the lowest values being in YE grapes at harvest) (Figure 4).

The knowledge of the carotenoid profile of grapes is linked to the type of norisoprenoid compound which will form in wine. β-ionone is derived from β-carotene oxidation, whereas 3-hydroxy-α-ionone and 3-oxo-α-ionol originate by lutein cleavage [11,12]. The factor score plot (accounting for 82.06% of total variance) from the PCA performed on the variables corresponding to the difference of the identified carotenoid level between véraison and harvest in both YE and C grape samples is shown in Figure 7. This clearly shows that YE-treated grapes of Negro Amaro and Primitivo were different from C samples along PC_1 and PC_2, respectively, and they were more correlated to carotenoid-derivative variables. Furthermore, NYE was mainly characterized by lutein, β-carotene, and violaxanthin derivatives (factor coordinates: −0.9535, −0.8657, and −0.8643, respectively, onto PC_1), while PYE was principally characterized by a higher concentration of (9′Z)-neoxanthin derivatives (factor coordinate: −0.8492 onto PC_2) (Figure 7).

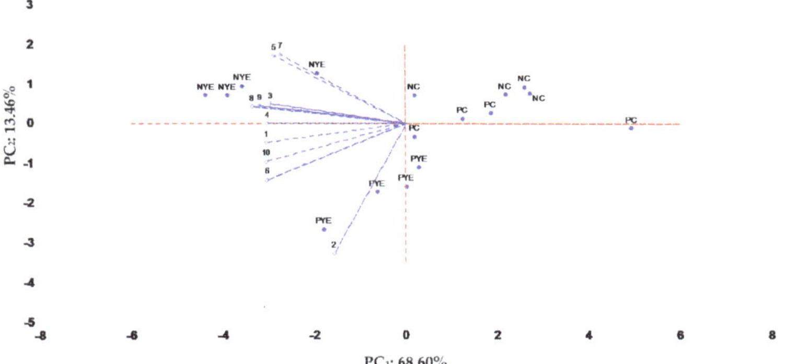

Figure 7. Principal component diagram of carotenoid derivatives in the two grape varieties (N: Negro Amaro; P: Primitivo) as affected by the treatment (C: control; YE: yeast extract). 1, violaxanthin derivative; 2, (9′Z)-neoxanthin derivative; 3, (8′S)-neochrome derivative; 4, 5,6-epoxylutein derivative; 5, luteoxanthin derivative; 6, lutein like structure derivative; 7, Z-lutein like structure derivative; 8, (allE)-lutein derivative; 9, (9Z)-lutein derivative; 10, β-carotene derivative.

4. Conclusions

Fifteen carotenoids, including (allE)-lutein and β-carotene together with their 9Z isomers and 5,6-/5,8-epoxyxanthophylls, were tentatively identified by matching UV-Vis characteristics, MS spectra, and elution order, and were accurately quantified by the used analytical method. As expected, a decrease of carotenoids during ripening was observed in both the two cultivars, but this was even more noticeable in grapes treated with YE. β-carotene and (allE)-lutein were present as the main compounds in all the analyzed samples, and the level of xanthophyll appeared to be more drastically reduced than the carotene level, especially in mature Negro Amaro and Primitivo treated grapes. Similarly, the elicitor treatment caused a more consistent decrease of violaxanthin, (9′Z)-neoxanthin, and 5,6-epoxylutein.

Thereby, because carotenoids serve as precursors of C_{13}-norisoprenoids, the YE treatment proved to be determinant in enhancing the aroma potential (ΔC) of both varieties up to four-fold compared to untreated grapes. Besides this, Negro Amaro and Primitivo were principally characterized by lutein, β-carotene, and violaxanthin derivatives, and (9′Z)-neoxanthin derivatives, respectively. However, further research is ongoing to confirm the real characteristics of these grapes in obtaining wines which are richer in sensorial impact C_{13} aroma compounds, such as β-ionone and β-damascenone.

Supplementary Materials: The following are available online at http://www.mdpi.com/2076-3417/10/10/3369/s1, Figure S1: Extraction yields of (**a**) Negro Amaro and (**b**) Primitivo samples, Table S1: Validation parameters, linearity, repeatability (r), LOD and LOQ for HPLC-DAD analyses.

Author Contributions: Conceptualization, P.C.; methodology, G.M., A.R.C. and L.T.; validation, P.C. and M.S.; formal analysis, M.S.; data curation, P.C.; writing—original draft preparation, P.C.; writing—review and editing, F.V. and F.A.T.-B.; visualization, F.B. All authors have read and agreed to the published version of the manuscript.

Funding: This research received no external funding.

Acknowledgments: We would like to thank LALLEMAND Italia S.p.A. for supplying the Lalvigne® MATURE formulation used in this work.

Conflicts of Interest: The authors declare no conflict of interest. LALLEMAND had no role in the design of the study; in the collection, analyses, or interpretation of data; in the writing of the manuscript, or in the decision to publish the results.

References

1.	Razungles, A.; Babic, I.; Sapis, J.C.; Bayonove, C.L. Particolar behavior of epoxy xanthophylls during veraison and maturation of grape. *J. Agric. Food Chem.* **1996**, *44*, 3821–3825. [CrossRef]

2.	Razungles, A.; Bayonove, C.L.; Cordonnier, R.E.; Baumes, R.L. Etude des Carotenoides du Raisin à Maturitè. *Vitis* **1987**, *26*, 183–191.

3.	Razungles, A.; Bayonove, C.L.; Cordonnier, R.E.; Sapis, J.C. Grape carotenoides: Changes during the maturation period and localization in mature berries. *Am. J. Enol. Vitic.* **1988**, *39*, 44–48.

4.	Crupi, P.; Milella, R.A.; Antonacci, D. Simultaneous HPLC-DAD-MS (ESI+) determination of structural and geometrical isomers of carotenoids in mature grapes. *J. Mass Spectrom.* **2010**, *45*, 971–980. [CrossRef] [PubMed]

5.	Mendes-Pinto, M.M.; Silva-Ferreira, A.C.; Caris-Veyrat, C.; Guedes de Pinho, P. Carotenoid, chlorophyll, and chlorophyll-derived compounds in grapes and port wines. *J. Agric. Food Chem.* **2005**, *53*, 10034–10041. [CrossRef] [PubMed]

6.	Silva-Ferreira, A.C.; Monteiro, J.; Oliveira, C.; Guedes de Pinho, P. Study of major aromatic compounds in port wines from carotenoid degradation. *Food Chem.* **2008**, *110*, 83–87. [CrossRef]

7.	Crupi, P.; Coletta, A.; Milella, R.A.; Palmisano, G.; Baiano, A.; La Notte, E.; Antonacci, D. Carotenoid and Chlorophyll derived compounds in some wine grapes grown in Apulian region. *J. Food Sci.* **2010**, *75*, S191–S198. [CrossRef]

8.	Azevedo-Meleiro, C.H.; Rodriguez-Amaya, D.B. Confirmation of the identity of the carotenoids of tropical fruits by HPLC-DAD and HPLC-MS. *J. Food Compos. Anal.* **2004**, *17*, 385–396. [CrossRef]

9.	Crupi, P.; Alba, V.; Masi, G.; Caputo, A.R.; Tarricone, L. Effect of two exogenous plant growth regulators on the color and quality parameters of seedless table grape berries. *Food Res. Int.* **2019**, *126*, 108667. [CrossRef]

10.	Baumes, R. Wine aroma precursors. In *Wine chemistry and biochemistry*; Moreno-Arribas, M.V., Polo, M.C., Eds.; Springer: New York, NY, USA, 2009; pp. 251–274.

11.	Crupi, P.; Coletta, A.; Antonacci, D. Analysis of carotenoids in grapes to predict norisoprenoid varietal aroma of wines from Apulia. *J. Agric. Food Chem.* **2010**, *58*, 9647–9656. [CrossRef]

12.	Baumes, R.; Wirth, J.; Bureau, S.; Gunata, Y.; Razungles, A. Biogeneration of C13-norisoprenoid compounds: Experiments supportive for an apo-carotenoid pathway in grapevines. *Anal. Chim Acta* **2002**, *458*, 3–14. [CrossRef]

13.	Oliveira, C.; Silva Ferreira, A.C.; Mendes Pinto, M.; Hogg, T.; Alves, F. Carotenoid compounds in grapes and their relationship to plant water status. *J. Agric. Food Chem.* **2003**, *51*, 5967–5971. [CrossRef] [PubMed]

14.	van Breemen, R.B. Liquid chromatography/mass spectrometry of carotenoids. *Pure App. Chem.* **1997**, *69*, 2061–2066. [CrossRef]

15.	Emenhiser, C.; Simunovic, N.; Sander, L.C.; Schwartz, S.J. Separation of geometrical carotenoid isomers in biological extracts using a polymeric C30 column in reversed-phase liquid chromatography. *J. Agric. Food Chem.* **1996**, *44*, 3887–3893. [CrossRef]

16. Ferrari, S. Biological elicitors of plant secondary metabolites: Mode of action and use in the production of nutraceutic. In *Bio-Farms for Nutraceuticals: Functional Food and Safety Control by Biosensors*; Giardi, M.T., Rea, G., Berra, B., Eds.; Landes Bioscience and Springer Science + Business Media LLC: New York, NY, USA, 2010; pp. 152–166.

17. Bektas, Y.; Eulgem, T. Synthetic plant defense elicitors. *Front. Plant Sci.* **2015**, *5*, 1–17. [CrossRef]

18. Portu, J.; López, R.; Baroja, E.; Santamaría, P.; Garde-Cerdán, T. Improvement of grape and wine phenolic content by foliar application to grapevine of three different elicitors: Methyl jasmonate, chitosan, and yeast extract. *Food Chem.* **2016**, *201*, 213–221. [CrossRef]

19. Šuklje, K.; Antalick, G.; Buica, A.; Coetzee, Z.A.; Brand, J.; Schmidtke, L.M.; Vivier, M.A. Inactive dry yeast application on grapes modify Sauvignon Blanc wine aroma. *Food Chem.* **2016**, *197*, 1073–1084. [CrossRef]

20. Gutiérrez-Gamboa, G.; Marín-San Román, S.; Jofré, V.; Rubio-Bretón, P.; Pérez-Álvarez, E.P.; Garde-Cerdán, T. Effects on chlorophyll and carotenoid contents in different grape varieties (Vitis vinifera L.) after nitrogen and elicitor foliar applications to the vineyard. *Food Chem.* **2018**, *269*, 380–386. [CrossRef]

21. Betz, M.; Schindler, C.; Schwender, J.; Lichtenthaler, H.K. Jasmonic acid induced changes in carotenoid levels and zeaxanthin cycle performance. In *Plant Lipid Metabolism*; Kader, J.C., Mazliak, P., Eds.; Springer: Dordrecht, The Netherlands, 1995; pp. 353–355.

22. Lu, Y.; Jiang, P.; Liu, S.; Gan, Q.; Cui, H.; Quin, S. Methyl jasmonate- or gibberellins A3-induced astaxanthin accumulation is associated with up-regulation of transcription of β-carotene ketolase genes (bkts) in microalga *Haematococcus pluvialis*. *Bioresour. Technol.* **2010**, *101*, 6468–6474. [CrossRef]

23. Office International de la Vigne et du Vin (OIV). *Recueil des Methodes Internationales d'Analyse des Vins et des Mouts*; Office International de la Vigne et du Vin: Paris, France, 1990.

24. Melendez-Martinez, A.J.; Britton, G.; Vicario, I.M.; Heredia, F.J. HPLC analysis of geometrical isomers of lutein epoxide isolated from dandelion (Taraxacum officinale F.Weber exWiggers). *Phytochemistry* **2006**, *67*, 771–777. [CrossRef]

25. Martì, M.P.; Busto, O.; Guash, J. Application of a headspace mass spectrometry system to the differentiation and classification of wines according to their origin, variety and ageing. *J. Chromatogr. A* **2004**, *1057*, 211–217. [CrossRef] [PubMed]

26. Oliver, J.; Palou, A. Chromatographic determination of carotenoids in foods. *J. Chromatogr. A* **2000**, *881*, 543–555. [CrossRef]

27. Bureau, S.M.; Razungles, A.J.; Baumes, R.L.; Bayonove, C.L. Effect of vine or bunch shading on the carotenoid composition in Vitis Vinifera L. berries. I. Syrah grapes. *Vitic. Enol. Sci.* **1998**, *53*, 64–71.

28. González-Gómez, D.; Lozano, M.; Fernández-León, M.F.; Bernalte, M.J.; Ayuso, M.C.; Rodríguez, A.B. Sweet cherry phytochemicals: Identification and characterization by HPLC-DAD/ESI-MS in six sweet cherry cultivars grown in Valle del Jerte (Spain). *J. Food Compos. Anal.* **2010**, *23*, 533–539. [CrossRef]

29. Oliveira, C.; Barbosa, A.; Silva Ferreira, A.C.; Guerra, J.; Guedes De Pinho, P. Carotenoid profile in grapes related to aromatic compounds in wines from Douro region. *J. Food Sci.* **2006**, *71*, S1–S7. [CrossRef]

30. Gómez-Plaza, E.; Mestre-Ortuño, L.; Ruiz García, Y.; Fernández-Fernández, J.I.; López-Roca, J.M. Effect of benzothiadiazole and methyl jasmonate on the volatile compound composition of Vitis vinifera L. Monastrell grapes and wines. *Am. J. Enol. Vitic.* **2012**, *63*, 394–401. [CrossRef]

31. Lucini, L.; Baccolo, G.; Rouphael, Y.; Colla, G.; Bavaresco, L.; Trevisan, M. Chitosan treatment elicited defence mechanisms, pentacyclic triterpenoids and stilbene accumulation in grape (*Vitis vinifera* L.) bunches. *Phytochemistry* **2018**, *156*, 1–8. [CrossRef]

32. Kogkou, C.; Chorti, E.; Kyraleou, M.; Kallithraka, S.; Koundouras, S.; Logan, G.; Kanakis, I.; Kotseridis, Y. Effects of foliar application of inactivated yeast on the phenolic composition of Vitis vinifera L. cv. Agiorgitiko grapes under different irrigation level. *Int. J. Wine Res.* **2017**, *9*, 23–33. [CrossRef]

33. Paladines-Quezada, D.F.; Fernández-Fernández, J.I.; Bautista-Ortín, A.B.; Gómez-Plaza, E.; Bleda-Sánchez, J.A.; Gil-Muñoz, R. Influence of the use of elicitors over the composition of cell wall grapes. In Proceedings of the In Vino Analytica Scientia 2017, Salamanca, Spain, 17–20 July 2017.

© 2020 by the authors. Licensee MDPI, Basel, Switzerland. This article is an open access article distributed under the terms and conditions of the Creative Commons Attribution (CC BY) license (http://creativecommons.org/licenses/by/4.0/).

MDPI

St. Alban-Anlage 66

4052 Basel

Switzerland

Tel. +41 61 683 77 34

Fax +41 61 302 89 18

www.mdpi.com

Applied Sciences Editorial Office

E-mail: applsci@mdpi.com

www.mdpi.com/journal/applsci

www.ingramcontent.com/pod-product-compliance
Lightning Source LLC
LaVergne TN
LVHW080556180726
843515LV00004BB/43